Dedicate to

My Grandchildren (joie de vivre)
Kartik   Kapil   Raghav   Shyam

My Daughters (motivating me constantly)
Meera   Poornima   Shaila

My Wife (pillar of strength)
Latha

Contents

Foreword . . . . *ix*
Preface . . . . *xi*

Chapter 1: Earthing of Electrical System – Fundamentals 1
1.0   Introduction 1
1.1   Distinction Between Grounding and Earthing 1
1.2   Earth as Conductor 2
1.3   Effect of Moisture on Soil Resistivity 4
1.4   Effect of Salt on Soil Resistivity 4
1.5   Effect of Temperature on Soil Resistivity 6
1.6   Soil Resistivity Measurement 6
1.7   Electrode Resistance to Earth 9
1.8   CT and Earth Electrode 10
1.9   Hemispherical Electrode 10
1.10 Pipe Electrode or Driven Rod 12
1.11 Strip or Horizontal Wire Electrode 14
1.12 Influencing Factors for Electrode Resistance 15
1.13 Plate Electrode 16
1.14 Parallel Electrodes 17
1.15 Resistance of Earthing Grid 21
1.16 Methods to Reduce Earth Grid Resistance 23
1.17 Measurement of Electrode Resistance 24
1.18 Electrode Sizing 26

Chapter 2: Earthing of Electrical System – EHV System 29
2.0   Introduction 29
2.1   Human Element 29
2.2   Hotline Operation 30
2.3   Tolerable Current 31
2.4   Ground Potential Rise 34
2.5   Current Discharged to Earth - I_G 35
2.6   Fault on Transmission Line 37

2.7	Fault on EHV cable	42
2.8	Step and Touch Potentials	45
2.9	Effect of Thin Layer of Crushed Rock	50
2.10	Allowable Touch and Step Potentials	52
2.11	Design Touch and Step Potentials	54
2.12	Influence of Cross Section	57
2.13	Summary of Switch Yard Ground Grid Design	58
2.14	Earthing of Special Areas	59
2.15	Sample Outputs from Software (CDEGS)	61
2.16	GIS Earthing	65

Appendix 2-1: Division of Current Between Ground Wire and Earth ... 69

2-1.1.	Induction	69
2-1.2.	Conduction	72

Chapter 3: Earthing of Electrical System – LV & MV System ... 73

3.0	Introduction	73
3.1	Earthing in LV system	73
3.2	Earthing in MV Resistance Grounded System	84
3.3	Earthing in MV Ungrounded System	85
3.4	Miscellaneous Topics	86

Chapter 4: Grounding of Electrical System – Ungrounded System ... 95

4.0	Introduction	95
4.1	Ungrounded vs Grounded system	95
4.2	Difference Between Neutral and Ground	95
4.3	Ungrounded System (Balanced Operation)	96
4.4	Ungrounded System (Fault Condition)	97
4.5	Advantages of Ungrounded System	100
4.6	Disadvantages of Ungrounded System	101
4.7	Ground Fault Detection in Ungrounded System	102
4.8	Ground Fault Detection in DC System	102
4.9	Ground Fault Detection in AC System	105
4.10	Ground Fault Detection on Ships	107
4.11	Single Phase PT Connection for Ground Fault Detection	108
4.12	Neutral Inversion and Ferro-Resonance	109

Chapter 5: Grounding of Electrical System – Grounded System ... 117

5.0	Introduction	117
5.1	Reasons for Grounding	117
5.2	Grounding Locations	117
5.3	Grounding Methods	119
5.4	Measure of Grounding Effectiveness	119
5.5	Reason for Choosing $K_F \geq 0.6$	121
5.6	Solidly Grounded System	121
5.7	Advantages of Solidly Grounded System	122

5.8	Disadvantages of Solidly Grounded System	122
5.9	Resistance Grounded System	126
5.10	Reactance Grounded System	131
5.11	Grounding the Bus	140
5.12	Variation of Voltage with Fault Current	147
5.13	Unearthed Grade (UE)/Earthed Grade (E) Cable Selection	149
5.14	Phase Voltage and Zero Sequence Voltage During Ground Fault	150
5.15	Earthing of Delta	151
5.16	CT Secondary Grounding	155

Appendix 5-1: Grounding Transformer Specification without Ambiguity ... **159**

5-1.0	Introduction	159
5-1.1	Case Study 1 (without NGR)	159
5-1.2	Case Study 2 (with NGR)	163

Appendix 5-2: NGT with/without NGR ... **167**

5-2.0	Introduction	167
5-2.1	System Data	167
5-2.2	Case 1 - NGT without NGR	168
5-2.3	Case 2 - NGT with NGR	169
5-2.4	Reason for choosing only NGT and not NGT & NGR	169

Appendix 5-3: Sizing of NGT ... **171**

5-3.0	Analysis	171
5-3.1	Open Delta Voltage	172
5-3.2	Spread Sheet Calculation	173
5-3.3	Example	174

Appendix 5-4: Sizing of NGR ... **179**

5-4.0	Introduction	179
5-4.1	Spread Sheet Calculation	180
5-4.2	Example:	180

Appendix 5-5: Zig Zag – Star Transformer for Auxiliary Supply in Switchyard ... **185**

5-5.0	Introduction	185
5-5.1	Load Connected Between Phases of Star Winding	185
5-5.2	Load Connected Between Phases and Neutral of Star Winding	186

Appendix 5-6: Zig Zag Grounding Transformer – Short Circuit Testing ... **187**

5-6.0	Introduction	187
5-6.1	Alternative 1	187
5-6.2	Alternative 2	188
5-6.3	Alternative 2 – Theory	188

Appendix 5-7: Analysis of (L-L) Fault on Tertiary Winding in Autotransformer ... **191**

Chapter 6: Grounding – Fairy Tales 195
6.0 Introduction 195
Questions 195
Answers 202

Appendix 6-1: Analysis of Tripping of GSU By Differential Protection for External Fault 215
6-1.0 Introduction 215
6-1.1 Data 215
6-1.2 Analysis 216

Chapter 7: Generator Neutral Grounding Practices 221
7.0 Introduction 221
7.1 Harmonic and Zero Sequence 221
7.2 Generators Connected to a Common Bus 224
7.3 NGR Common to All the Units 226
7.4 Individual NGR for the Units 227
7.5 Zig Zag Grounding Transformer Common to All the Units 228
7.6 LV Generators Grounding 229
7.7 MV Generators Grounding 231
7.8 High Resistance Grounding 231
7.9 Low Resistance Grounding 235
7.10 Sensitivity of Ground Fault Protection 236
7.11 Hybrid Grounding 237
7.12 Grounding Mix-up 238

References 241

Foreword

Electric Energy (or Electricity, as commonly called) is being used for human mankind and economic development for almost 140 years. It is an extremely dangerous quantity. Scientists and engineers have been dealing with this and how to protect human and animal life and property while reaping the benefits of its enormous power. Grounding (or earthing) is one of the main defenses against hazardous electric shocks and protection of equipment.

Personnel safety is the most important consideration in all design and applications in electric power and energy systems. Grounding is one of the key debatable issues. Subject of grounding quite often is misunderstood and misapplied. People argued for and against it. Improper applications may cause fatal damages, loss of property and production. It is very important for all engineers and technical personnel to understand this subject.

This book is extremely timely and helpful for engineers to understand the underlining theory, assumptions and applications of proper grounding (or earthing) system design. It is written for all educators, practicing professionals at all level and our future workforce. It is written lucidly in a very simple style and understandable by all in our profession. The example problems are all practical and not found in any traditional textbooks, came from the author's 40 years of experience in the industry. The illustrations, sketches, pictures, drawings, phasor diagrams, tables and charts are done exceptionally well for clear understanding of the subject.

Yes, the market is changing how electricity will be delivered for the 21st century and beyond but the basic principles of electric power generation, transmission and distribution, and customer usage, the major equipment we use, the major infrastructure of the grid will not change as long as we need electricity. Safety will be the most important aspect in all applications and grounding will play an important role.

I have no doubt, this book will be used for years to come by hundreds and thousands of aspiring electric power and energy engineers all over the world. Every students, educators and practicing professionals (young and old) of electric power industry should have this book on their bookshelf for ready reference.

It is really a shame that this very important subject is not taught in academia except for a few lectures. This could be an extremely useful textbook.

This book is a testament to Dr. Rajamani's dedication to educate the future electrical power engineers who would take care of the world's electricity market for the 10 billion population by 2050. It is an exceptional effort and lifetime commitment by Dr. Rajamani. I applaud his gallant effort, sacrifices and time commitment to write this book for our future generations of electric power and energy engineers and sustainability for our planet.

Pankaj K (PK) Sen, Ph.D, Fellow IEEE, Professor
Professional Engineer
Colorado School of Mines
Golden, Colorado 80401, USA

Preface

After spending almost four decades in electrical industry, I accumulated 'enough information'. Also with the advent of powerful search engines on the internet, 'information is available on tap'. However this curse of information overload has lulled the design and practicing engineers into searching for readymade solutions. Also with the availability of easy to use power engineering software tools, design and analysis have been reduced to input – output exercise.

I tried to address the issue by conducting in depth workshops on various aspects of power engineering to design and field engineers. The aim was to attain the golden mean between too much scholarly material with lot of math and mundane project engineering. In this context I wrote a series of articles in IEEMA journal from 1999 onwards targeted towards field engineers. The feedback was very encouraging. I am indeed thankful to IEEMA for offering this platform for so long.

I thought time is ripe now to consolidate my previous work with additional supporting material. The outcome is the "Application Guide for Power Engineers". This will be in several parts to cover major topics in power engineering. This book is Part 1 and covers the most important topic of 'Earthing and Grounding of Electrical Systems'. A balanced mix of theory and practice is presented. The reader can quickly understand the basics and essential tips required during design and execution. I tried in my workshops explaining the concept of grounding using a series of questions and answers progressively increasing in complexity. I called these fairy tales. At the end of the tales, concept of grounding became bleedingly obvious. I have incorporated the same in Chapter 6 of this book.

This book is recommended for use by all design engineers before they embark on software simulations to judge the appropriateness of output from practical point of view. This book can be used by field engineers to understand the basics necessary to resolve site issues and augment/change existing system with confidence. This book can be used as a spring board to understand basic concepts and practices adopted at site before wading into Codes and Standards.

Graduate students will tremendously benefit from this book as it combines theory with lot of practical examples. The aspiring student will acquire necessary skill upfront needed by industry.

This book is outcome of decades of experience. I have interacted with so many professionals in the field who have enriched my knowledge and my thanks to all of them. I wish to thank D Guha for his pointed criticism, and prodding me to logical conclusion 'where the rubber meets the road'.

Harish Mehta gave me the initial spark to organize workshops for field engineers to explain the practical aspects supported by theory. He taught me the art of discarding the superficial and concentrating on the essential.

I wish to thank Bina Mitra who has been my coauthor for so many of my articles. Many of the examples presented in Part 1 as well to be presented in future Parts are outcome from her field experience in different sites. She had been my intellectual prod who executed many of my ideas at site.

I am indebted to Isaac Izraeli for crisp explanation on concept of earth electrode resistance.

I acknowledge the inputs received on practical aspects copiously quoted in the book from following persons: Sandy Murdoch, Cajetan Pinto, Rajesh Vadangekar, Tukaram Talande, Jafar Khan, Sandeep Godbole, Mohan Waingankar, Ashutosh Pailwan, Vini Vazhappully, M V Kini, Abhijit Mandal and Gouni Reddy.

Special thanks to Sonu Karekar and Amol Salunkhe who have done simulation studies using PSSE, NEPLAN and PSCAD and verified the results of examples given in the book.

Results of earthing grid design for large power plants done using CDEGS software were provided by Bodhlal Prasad.

Deliberations with Prof. S V Kulkarni helped the author in understanding the nuances of Zig Zag connection.

I owe my intellectual debt to Prof. M V Hariharan. Discussions with him always bring clarity of thought.

Sangeetha (Kshiteeja Gamre) worked tirelessly tolerating my demand for endless revisions in preparing the final manuscript and her secretarial assistance is gratefully acknowledged.

I wish to thank the management of Adani Electricity Mumbai Ltd (formerly Reliance Infrastructure Ltd) for their continuous support both in my professional life as well as personal life. The liberal attitude of the management in allowing me to 'experiment' with new ideas and its unstinted support in providing material and manpower is acknowledged.

Special thanks to Sandeep Sarma for coordinating with publishers and getting the final product out.

Before I conclude, I salute Dr. Ajit Menon. He has restored my health and instilled confidence to continue my professional work.

Earthing of Electrical System – Fundamentals

Chapter 1

1.0 INTRODUCTION

Earthing of electrical system is a unique topic. It is very important from the view of safety and protection. But there is a lot of myth in this area carried from the past. The objective here is to clarify and clear these myths and bring out the salient features that truly improve the earthing system performance. In this Chapter basic concepts are introduced [3]. Application to EHV switchyards and MV & LV systems are covered in subsequent chapters.

1.1 DISTINCTION BETWEEN GROUNDING AND EARTHING

Grounding implies connection of *current carrying* parts to ground. It is mostly either generator or transformer neutral. Hence it is popularly called 'neutral grounding'. Grounding is for *equipment safety*. Details of grounding practices will be covered in Chapters 4 to 7.

Earthing implies connection of *non-current carrying* parts to ground like metallic enclosures (Figure 1.1).

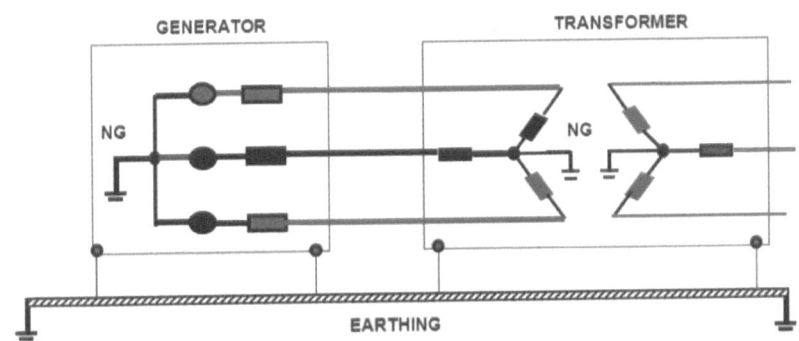

Figure 1.1: Neutral Grounding and Earthing

Earthing is for *human safety*. Under balanced operating conditions of power systems, earthing system does not play any role. But during ground fault condition, it enables the ground fault current to return to the source without endangering human safety.

The above definition is not rigid. Many times terms earthing and grounding are used interchangeably. By context the terms are interpreted either as neutral earthing or body earthing.

Four constituents under consideration are shown in Figure 1.2. First is the Power System which is the active component. Second is the earthing system to provide a return path for current to return to source. Third is the human being for whose safety the earthing system is designed. Fourth is 'earth' which is the medium for carrying return fault current. It is not mandatory to have fourth element 'earth' for carrying fault current back to source as in case of ships or oil platforms.

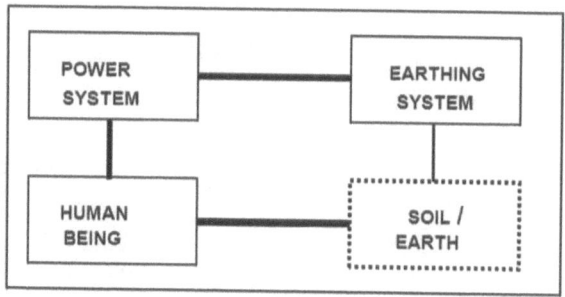

Figure 1.2: Four constituents of system earthing

1.2 EARTH AS CONDUCTOR

1.2.1 Resistivity of Element

Resistance of element is given by:

$$R = \frac{\rho L}{A} = \Omega \tag{1.1}$$

L: Length of element (M)
A: Cross sectional area (M²)
ρ: Resistivity of element

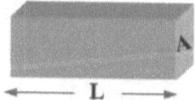

From Eqn (1.1)

$$\rho = \frac{RA}{L} = \frac{\Omega M^2}{M} = \Omega M \qquad (1.2)$$

Hence, the unit of dimension for resistivity of any element including soil is ΩM. It is the resistance offered between two faces by 1M cube.

1.2.2 Mother Earth

Myth: 'Mother Earth' is a good equipment grounding conductor
Reality: Earth is one of the worst equipment grounding conductors.

Contrary to popular perception, 'mother earth' is a bad conductor. Refer Figure 1.3. Resistivity (ρ) of earth is typically $100\Omega M$. In comparison, Resistivity (ρ) of copper is 1.7×10^{-8} ΩM and GI is 10^{-7} ΩM. Taking 25x4 mm Cu as reference, to obtain the same resistance, the size of GI will be 65x10 mm. The corresponding figure for earth is 800x800 meters (158 acres)! This comparison clearly shows that wherever possible, metallic conductor is a preferred alternative to earth to bring the fault current back to the source. In Chapters 1 to 3, the advantage of using metal as a conductor for taking the current back to source (e.g. armour, metallic sheath, ground wire, bonding conductor, etc) instead of depending on 'mother earth' will be brought out. However mother earth can offer a very large area to compensate for the high resistivity ρ.

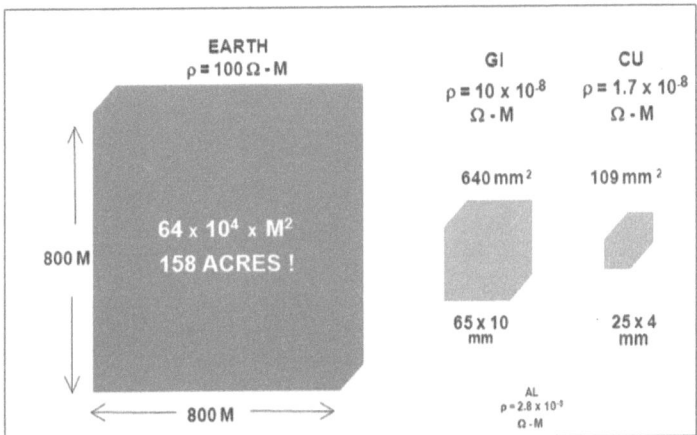

Figure 1.3: Comparison of Resistivity among Earth, GI and Copper

1.3 EFFECT OF MOISTURE ON SOIL RESISTIVITY

One M^3 of soil weighs approximately 1500 Kg. If moisture content is less than 10% of soil weight, there is steep increase in ρ. Addition of water increases moisture content in soil with corresponding reduction in ρ.

If moisture content is more than 25%, the soil is almost saturated with water and there is not much reduction in ρ thereafter. Refer Figure 1.4.

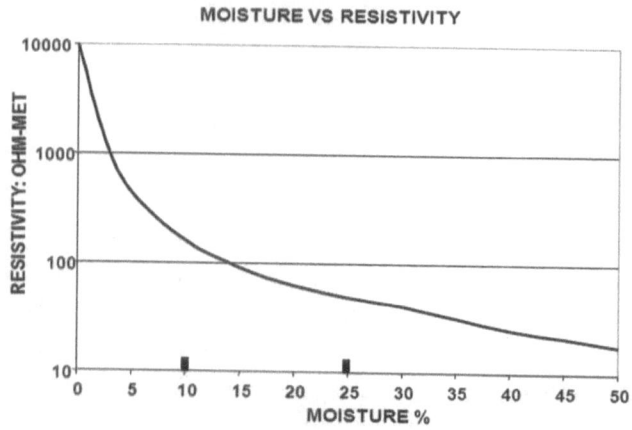

Figure 1.4: Moisture vs Soil Resistivity

1.4 EFFECT OF SALT ON SOIL RESISTIVITY

Even minute amount of salt results in sharp decrease in ρ. Refer Figure 1.5. This decrease is possible with only sufficient moisture content. Addition of salt to dry soil will not result in much improvement.

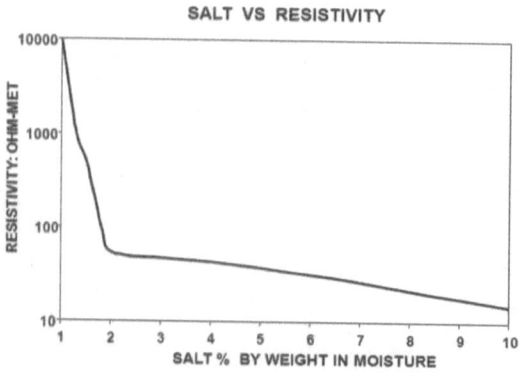

Figure 1.5: Effect of salt on soli resistivity

Generally used salts are Sodium chloride (Common Salt), Copper Sulphate, Calcium Chloride and Magnesium Sulphate. Addition of salt in soil results in corrosion of metal (electrode) embedded in soil. When sizing electrodes, the size is increased to account for corrosion. The corrosion allowance is generally taken as 15%.

To mitigate the effect of corrosion, instead of salt, alternative substances have been tried for treating the soil to reduce ρ. Bentonite clay (chemical name is sodium montmorillonite) is a popular alternative widely used. It is very benign on metals. Resistivity of Bentonite is less than $5\Omega M$. The electrode (say 50 mm dia, 3M long GI) is surrounded by a mix of Bentonite and native soil. The typical mix is 1:5 (Bentonite to soil) by weight. However abundant water supply shall be ensured when soil is treated with Bentonite clay to achieve desired reduction in resistivity. Carbon based backfill compounds are also available and is claimed to be as effective as Bentonite.

The soil classification based on corrosion intensity is given below

ρ of Virgin Soil (Ω - Met)	<25	25 <ρ< 50	50 <ρ< 100	ρ>100
Corrosion Intensity	Severe	Moderate	Mild	Very Mild

Table 1.1

Performance Over Time

Effect of artificial treatment of soil with salt is given in Figure 1.6. After treatment there is decrease in ρ. But with time, there is a gradual increase in ρ as salt is washed away by continual water seepage. Typically re-treatment of soil is done every three to five years.

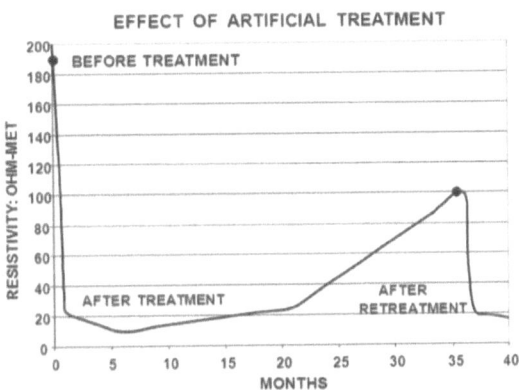

Figure 1.6: Effect of Artificial Treatment

1.5 EFFECT OF TEMPERATURE ON SOIL RESISTIVITY

The discussion here is relevant only in very cold climates. ρ decreases with increase in temperature. In summer ρ is less and in winter ρ is more. Performance of soil is better (like battery) at higher temperature. Effect of temperature on ρ is not serious until freezing point is approached. Near 0°C, ρ abruptly rises to a very high value. Refer Figure 1.7. Even though the atmospheric temperature may be near zero, just a few meters below the ground, the ground temperature is much higher. This is the reason earth electrodes are buried deep down the ground in cold climates.

The converse is also true. Even if atmospheric temperature is high, the soil temperature one meter below ground can be 3°C to 5°C lower.

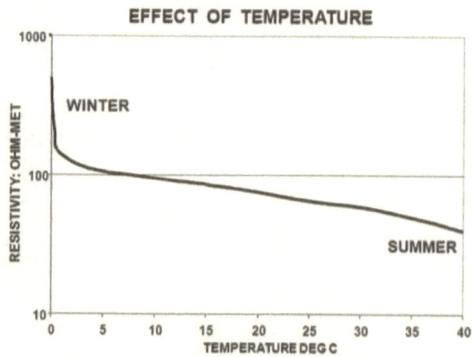

Figure 1.7: Effect of Temperature

1.6 SOIL RESISTIVITY MEASUREMENT

The connection diagram for Wenner's Four Probe method is shown in Figure 1.8. The basic principle is to inject the current and measure the voltage (not vice versa). It is in line with ultimate requirement where the fault current is injected into the ground and resulting Ground Potential Rise (GPR) is to be estimated. GPR is defined in Chapter 2.

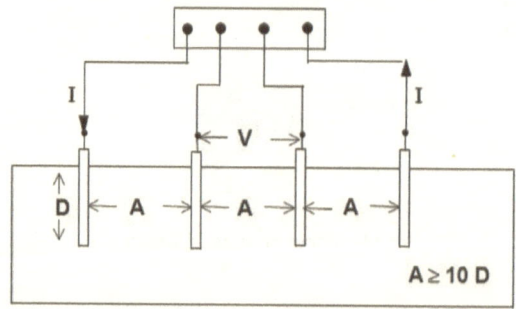

Figure 1.8: Four Probe Method

The current (I) is injected between two outer electrodes. Resulting voltage (V) is measured between two inner electrodes. D is the length of test electrodes (0.5M to 3M) and A is the distance between the electrodes. A is varied from 2 to 20M. ρ is evaluated from following formula:

R = V/I

$$\rho = \frac{4\pi A R}{1 + \frac{2A}{\sqrt{A^2 + 4D^2}} - \frac{2A}{\sqrt{4A^2 + 4D^2}}} \tag{1.3}$$

Since D << A

ρ ≅ 2πAR

Measured resistivity for spacing A represents apparent soil resistivity to depth of A. Refer Figure 1.9. Make measurements with different spacing. If ρ rapidly increases with spacing A, it Indicates underlying stratum is rock and it is difficult to install earth electrodes to great depths.

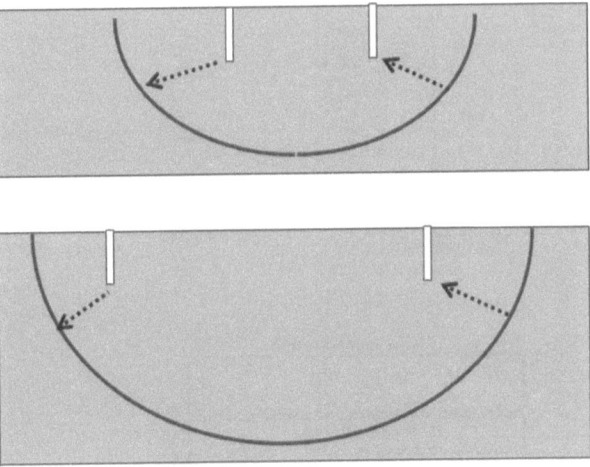

Figure 1.9: Spacing vs Depth

Soil resistivity measurements shall be made before any bare conductors or pipe lines are laid in the area of Test; otherwise the results may be corrupted. Usually this is one of the first activities taken up during the start of any project.

The measurement format is as follows. Readings are taken in four radials as shown in Figure 1.10. Along each radial, measurements are taken at different spacing.

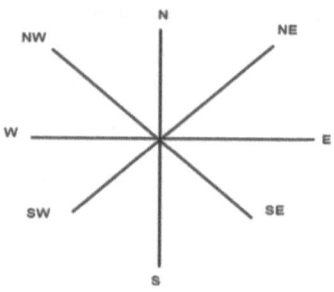

Spacing: 1, 2, 5, 10, 20, 50 M

Figure 1.10: Measurement Radials

Based on measurements, using software like CDEGS, soil resistivity at different layers can be evaluated as shown in Figure 1.11. In this example, soil resistivity is lowest between 1.5M and 3.7M. It gives a clue that for economical design, earthing grid be installed 1.5M below ground. More discussions follow in Sec 2.15.2.

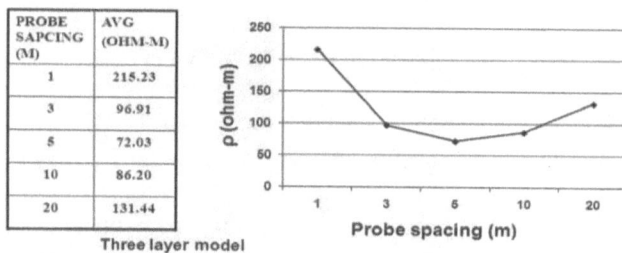

PROBE SAPCING (M)	AVG (OHM-M)
1	215.23
3	96.91
5	72.03
10	86.20
20	131.44

Three layer model

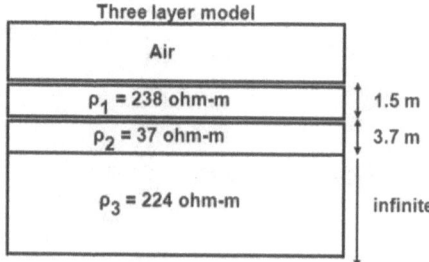

Figure 1.11: Multilayer Model

1.7 ELECTRODE RESISTANCE TO EARTH

It is the resistance between metal of electrode and general mass of earth. It is the resistance between specific electrode and imaginary electrode of zero resistance placed at infinite distance. Refer Figure 1.12. It will be shown later that 95% of resistance is contributed by soil within a few meters of electrode.

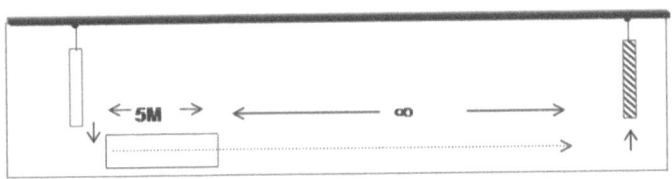

Figure 1.12: Electrode Resistance to Earth

A lot of conceptual confusion arises as the practicing engineer extrapolates 'conventional ohms law resistance' to electrode resistance. It does *not* reflect the situation where you apply a voltage across the electrode and measure the current and the resulting resistance is less than, say, 1Ω. For electrode resistance to earth, current is injected into the earth by electrode and the electric field travels through the earth. Current flows through a series of hemispherical shells of earth of continuously increasing cross section. Refer Figure 1.13. The voltage appears at certain distance from electrode and the resulting impedance is 'electrode resistance to earth'.

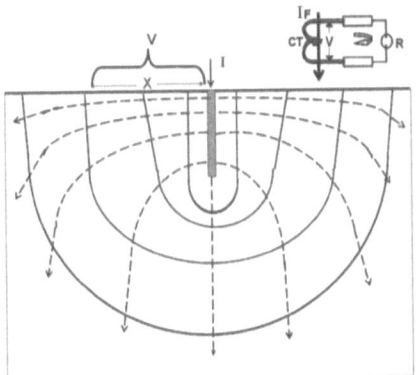

Figure 1.13: Resistance Area of driven earth rod

In majority of power system applications, current is injected into earth at the point of fault for a very short duration. In High Voltage Distribution System (HVDS) applications, load current (single phase) is injected into earth on a continuous basis if earth return is used.

1.8 CT AND EARTH ELECTRODE

Conceptually CT and Earth Electrode are same. In CT, the initiating quantity is primary current and the voltage appears on CT Secondary to drive current through connected burden. Refer Figure 1.13.

In Earth Electrode, the initiating quantity is ground fault current. This sets up electric field that penetrates spherical shells of increasing size. Voltage appears at any distance from electrode. In soil resistivity measurement or Earth Electrode resistance measurement, current is injected and resultant voltage is measured.

1.9 HEMISPHERICAL ELECTRODE

Consider a hemispherical electrode used for injecting the current. Refer Figure 1.14. Surface area of hemispherical earth at a distance X from electrode is $2\pi X^2$ (surface area of full sphere of radius X is $4\pi X^2$). The resistance offered by hemispherical earth of elemental thickness dX is given by

$R = \rho L/A = \rho\, dX/2\pi X^2$

Resistance offered by hemispherical earth of radius X is given by:

$$R_x = \int_0^x \frac{\rho dx}{2\pi X^2}$$

(1.4)

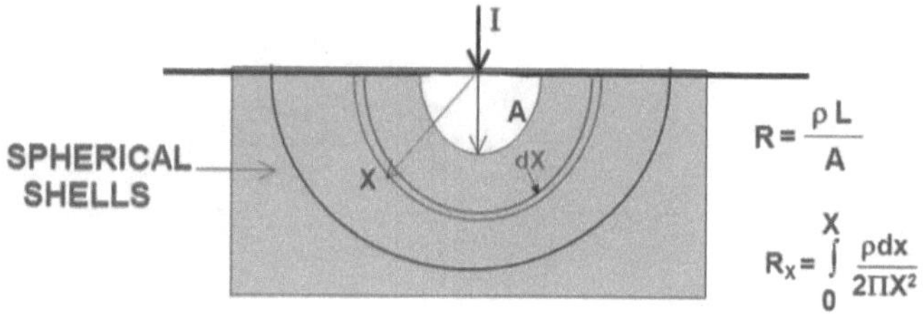

Figure 1.14: Hemispherical Electrode

Using Eqn (1.4), the resistance as a function of distance from electrode is shown in Figure 1.15. The most striking aspect of this curve is that almost 95% of final resistance is contributed by soil within 5 meters from electrode.

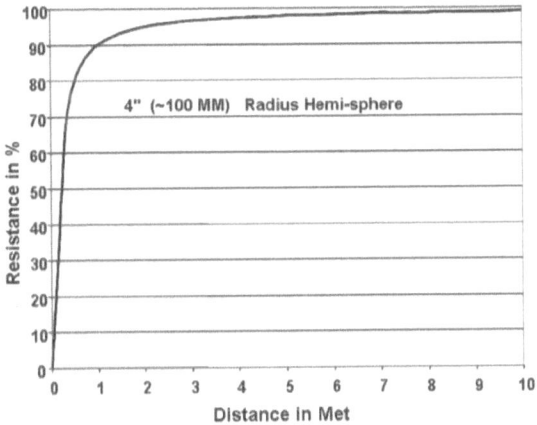

Figure 1.15: Resistance Vs Distance

The underlying concept is depicted in Figure 1.16. Resistance offered by elemental earth of same thickness decreases rapidly with distance from electrode.

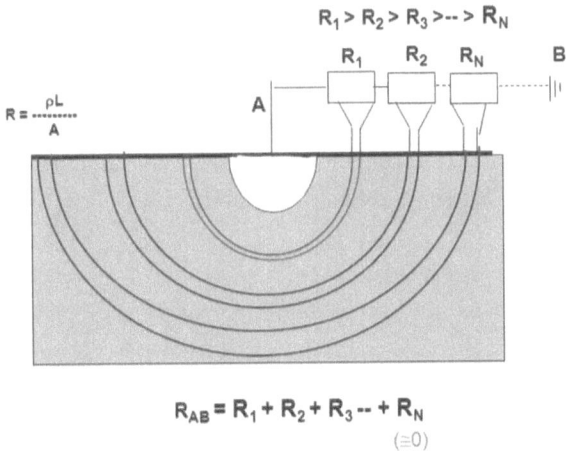

Figure 1.16: Resistance offered by earth of same elemental thickness

The percentage contribution out of total resistance as a function of distance from electrode is shown in Figure 1.17.

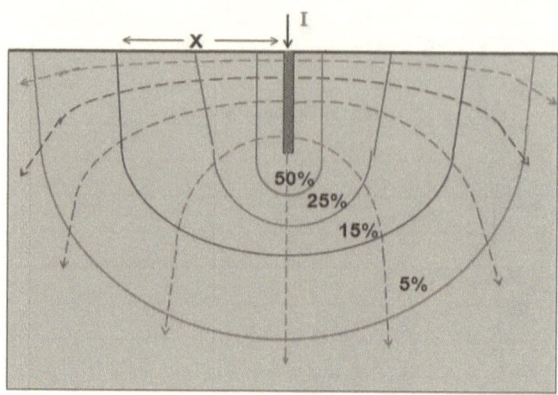

Figure 1.17: Percentage contributions to total resistance

Consider two substations A and B 100 KM apart with respective earth grids. Assume current is discharged at A. Only the soil within first five meters from A offers substantial resistance. The resistance offered by earth subsequently to reach B is very minimal. This confirms our practice of treating earth pits. By treating the soil *locally* around the electrode we are able to reduce electrode resistance, as the influence of earth away from electrode is minimal.

Another way to appreciate 'local effect' is based on the fact that earth with its huge mass offers almost ideal equipotential surface. A very large charge is required to change earth potential everywhere. Since the capacitance (C) of earth is extremely large, change in voltage is insignificant even if current (or charge) is dumped to the earth as $V = Q/C$. Any disturbance due to current injection is felt only locally.

1.10 PIPE ELECTRODE OR DRIVEN ROD

Pipe electrode (Figure 1.18) driven vertically is one of the most widely used electrodes in earthing grids. The resistance area for this case is shown in Figure 1.19. At sufficient distance from electrode, the electric field encounters shells that are almost hemispherical. Hence the conclusions drawn for hemispherical electrode are valid here also.

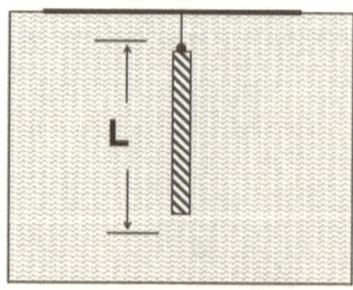

Figure 1.18: Pipe Electrodes

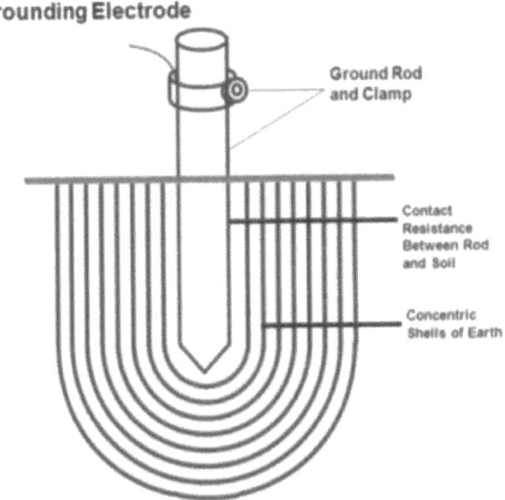

Figure 1.19: Resistance Area

The resistance of pipe electrode is given by:

$$R = \left(\frac{\rho}{2\pi L}\right)\left[ln\left(\frac{8L}{\Phi \times 2.7183}\right)\right] \tag{1.5}$$

L: Length in Met; Φ: Diameter in Met
Using Eqn (1.5), variation of resistance with length and diameter is given in Figure 1.20

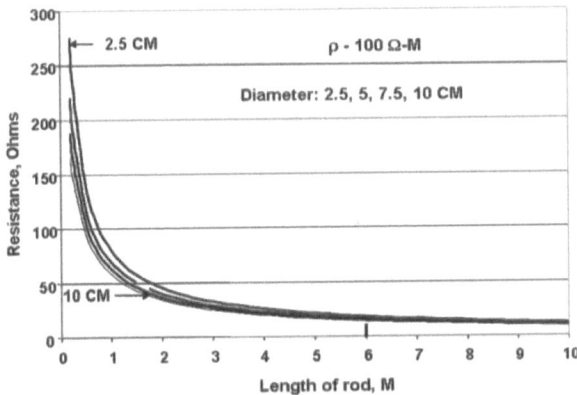

Figure 1.20: Resistance Vs length of pipe electrode

The length of electrode has major impact while the diameter has very minor influence. As an example, consider the case for length of 6M.

For Φ = 2.5 cm, R = 17.4Ω
For Φ = 10 cm, R = 13.7Ω
For 300% increase in diameter, resistance decreases by only 20%.

Typically length of electrode is less than 4M and at the most 6M. Beyond 6M, resistance value of individual electrode is almost constant.

1.11 STRIP OR HORIZONTAL WIRE ELECTRODE

The earth mats of EHV switchyards extensively use strip or rod electrodes (Figure 1.21). Typical layout of earthing grid formed with strip electrodes is shown in Figure 1.36.

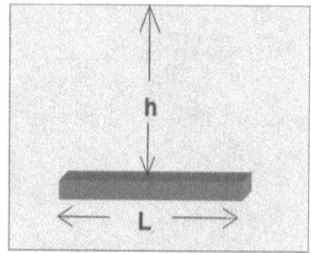

Figure 1.21: Strip electrode

The resistance of strip electrode is given by the well known Ryder's formula:

$$R = \frac{\rho}{2\pi L}\left[ln\left(\frac{8L}{T}\right) + ln\left(\frac{L}{h}\right) - 2 + \left(\frac{2h}{L}\right) - \left(\frac{h^2}{L^2}\right)\right] \tag{1.6}$$

L: Length in Met
h: Depth in Met
T: Width in Met (for strip)
 : 2 x diameter in Met (for wire or rod)

Using Eqn (1.6), variation of resistance with length and diameter is given in Figure 1.22.

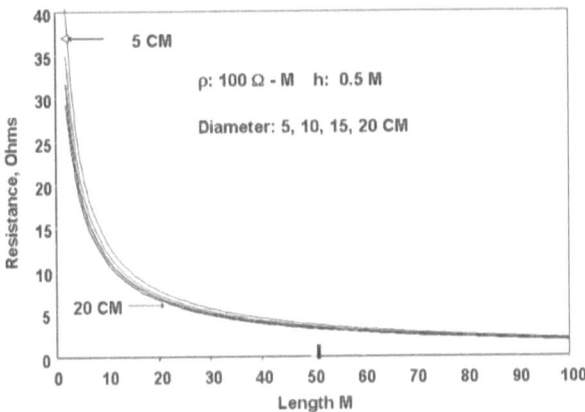

Figure 1.22: Resistance vs length

As in pipe electrode the length of electrode has major impact while the diameter has very minor influence. As an example, consider the case for length of 50M.

For Φ = 5 cm, R = 3.7Ω
For Φ = 20 cm, R = 3.2Ω
For 300% increase in diameter, resistance decreases by only 14%.

1.11.1 Circular vs Rectangular Cross Section

When applying Ryder formula (Eqn 1.6), Strip with width 'W' can be approximated to round conductor with diameter of 'W/2'. Refer Figure 1.23.

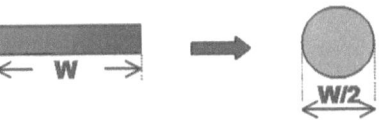

Figure 1.23: Strip vs Round Electrode

1.12 INFLUENCING FACTORS FOR ELECTRODE RESISTANCE

The major factor is the *length*. Diameter/width (cross section) has very minor influence. The other interesting observation is that the electrode resistance is not dependent on type of electrode material like Cu, Al or GI. It is a function of physical dimensions (mainly length) and not on physical properties. A horizontal earth strip of 75x10 mm Cu and 45x10 mm GI, both of same length will offer almost same electrode resistance. In conventional ohms law resistance, increased cross sectional area or use of Cu would signify smaller resistance but they are irrelevant as regards electrode resistance to earth is concerned.

Finally, the soil resistivity (ρ) has a linear impact.

Summarizing, resistance of electrode to earth is proportional to maximum dimension (length). It is not much influenced by minor dimensions like diameter or width. It is independent of material. Refer Figure 1.24.

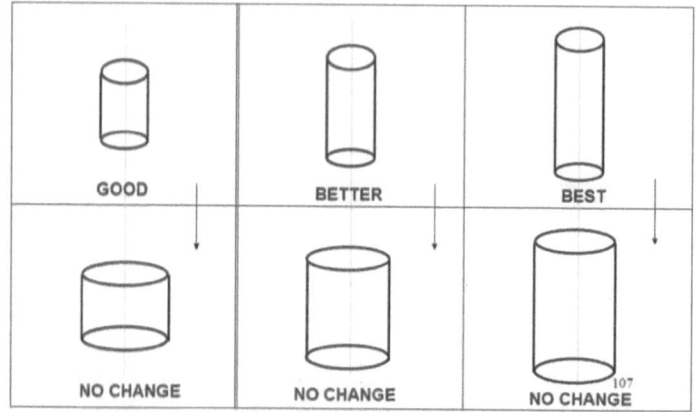

Figure 1.24: Electrode resistance vs length and cross section

1.13 PLATE ELECTRODE

In early days only plate electrodes were used. It was presumed that to get low electrode resistance to earth, surface area should be large (again the extrapolation of conventional ohms law concept). This fallacy has persisted for a long time.

In Figure 1.25 one electrode which is a solid plate and the other an annular ring of 5 cm thickness but both of them with same radius of 50 cm are shown. Calculations show that resistance to earth in both cases is 29.2Ω. Even an annular ring of 1 cm thickness with radius of 70 cm will have the same resistance to earth (29.2Ω).

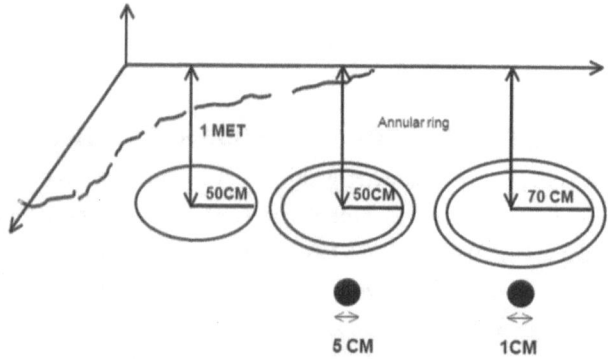

Figure 1.25: Resistance to earth - Plate Electrode

In Figure 1.26 plate electrode and strip electrode are shown having same volume or material. But resistance to earth of plate electrode is almost three times that of strip electrode. Intuitively it can be seen that linear dimension of plate electrode is 4M (perimeter) while that of strip electrode is 13M thus affirming the hypothesis that electrode resistance is dominated by 'length' of electrode buried. Thus it is concluded that plate electrode is very inefficient. It is rarely used in modern times.

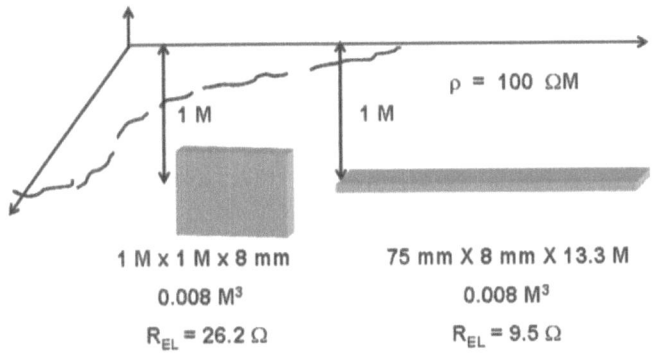

Figure 1.26: Resistance to earth of plate/strip electrode

1.14 PARALLEL ELECTRODES

1.14.1 Driven Rods or Pipe Electrodes in Parallel (Two Numbers)

To obtain low effective earth grid resistance, electrodes are connected in parallel. If resistance to earth of one electrode is 2Ω, the common perception is that effective resistance will be 1Ω if two such electrodes are connected in parallel (Figure 1.27). This is again due to our extrapolation of 'conventional ohms law' concept. Theoretically, the effective resistance will be half of 2Ω *provided the separation distance between electrodes is adequate.*

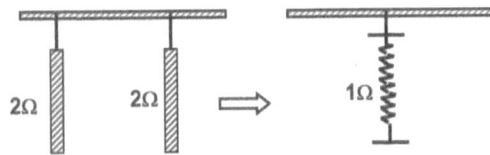

Figure 1.27: Parallel Electrodes

For discharging the electric field effectively, each electrode needs exclusive soil below it. If the rods are too close, resistance area of one electrode will interfere with that from other and expected gain is not realized. Refer Figure 1.28.

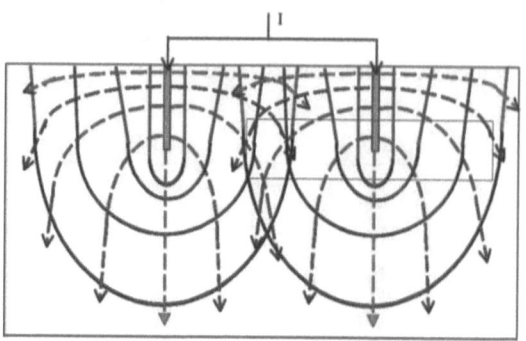

Figure 1.28: Overlapping resistance areas of two earth rods

The relationship between percentage effective resistance and separation distance is shown in Figure 1.29.

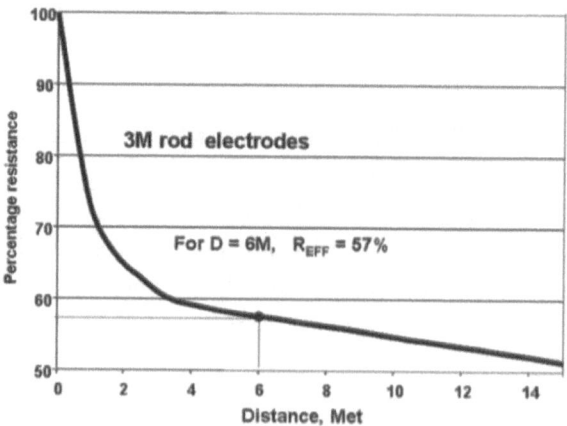

Figure 1.29: Effective Resistance Vs Separation distance (2 rods)

As a rule of thumb, if the rod length is L, minimum separation distance shall be 2L (Figure 1.30) wherever practicable. Effective resistance will be 50% only for very large separation distance,

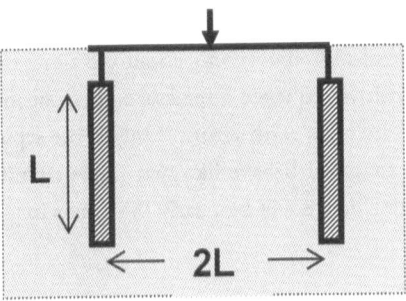

Figure 1.30: Separation distance

If equipment body or neutral is to be connected to two independent earth electrodes, it is preferable to locate the two electrodes on opposite sides of equipment to achieve large separation distance.

We will now detour a bit to review the sequence impedance of transmission line. Consider a Double Circuit EHV line with panther conductor (Figure 1.31).

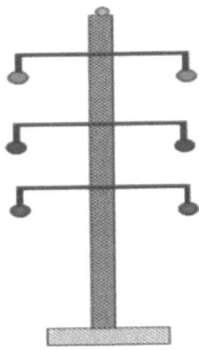

Figure 1.31: D/C Line

The following values (sequence impedances) are obtained from a line parameter evaluation program considering only one circuit and both the circuits:

	S/C	D/C
Z_{POS} (Ω/KM)	0.15 + j 0.41	0.08 + j 0.22
Z_{ZERO} (Ω/KM)	0.37 + j 1.29	0.29 + j 1.04

Table 1.2

Positive sequence impedance of D/C line is almost 0.5 times S/C line as expected. But Zero sequence impedance of D/C line is only about 0.8 times of S/C line. The intuitive rational for this is that positive sequence impedance does not involve earth return while zero sequence impedance involves earth return. Only if the separation distance between the two circuits is large, they will behave like two single circuit lines and the resulting effective zero sequence impedance will be nearly 0.5 times for S/C line. This is seldom achieved in practice.

1.14.2 Driven Rods or Pipe Electrodes in Parallel (Three Numbers)

The electrodes are placed at the corners of equilateral triangle (Figure 1.32). The relationship between percentage effective resistance and separation distance (sides of triangle) is shown in Figure 1.33. If the rod length is L, separation distance shall be at least 2L. Effective resistance will be 33% only for very large separation distance,

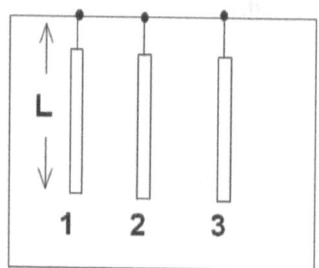
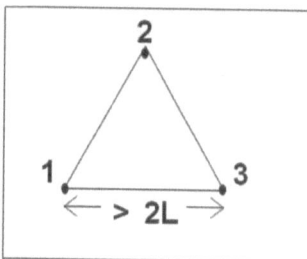

Figure 1.32: Three electrodes in parallel

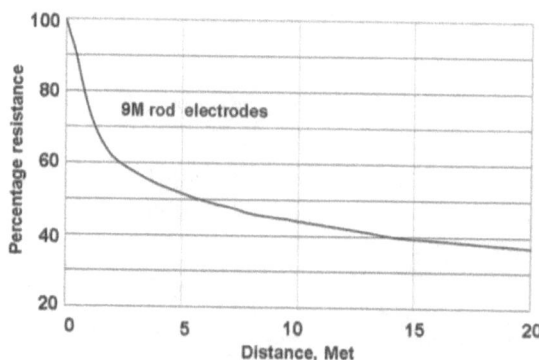

Figure 1.33: Effective resistance Vs Separation Distance (3 rods)

It is pertinent to bring out an analogy with planting of saplings by farmer which he does with instinct (Figure 1.34). The saplings are well spaced out so that each one has exclusive soil below for spreading its roots. If sapling are planted too close to each other, the roots of each sapling will compete for limited soil space below resulting in poor produce of low quality.

Figure 1.34: Sapling Separation

1.15 RESISTANCE OF EARTHING GRID

In previous sections, resistance to earth of individual electrodes was dealt with. In this section we will deal with earthing grid as a whole formed by a mesh of horizontal strip electrodes and vertical rod electrodes as in EHV switchyard (Figure 1.35).

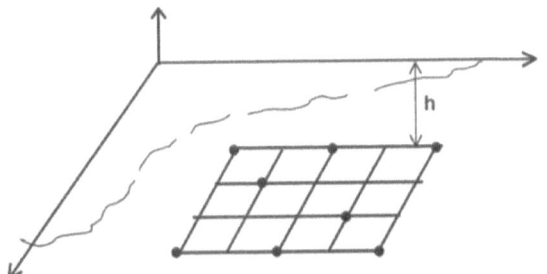

Figure 1.35: Switchyard Earthing grid

The resistance to earth of the entire grid is given by the famous Sverak formula:

$$C_1 = \frac{1}{L}; \quad C_2 = \frac{1}{\sqrt{(20A)}}; \quad C_3 = 1 + h\sqrt{\left(\frac{20}{A}\right)} \quad (1.7)$$

$$R_G = \rho\left[C_1 + C_2\left\{1 + \left(\frac{1}{C_3}\right)\right\}\right] \quad (1.8)$$

h: Depth of grid, Meter
A: Area of earthing grid, M²
L: Total length of buried conductor including vertical electrodes in Meter
ρ: Soil Resistivity in ΩM

Consider the example shown in Figure 1.36.

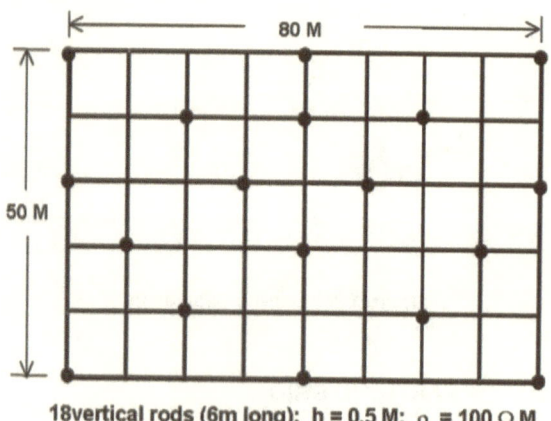

18vertical rods (6m long); h = 0.5 M; ρ = 100 Ω M

Figure 1.36: Rectangular Earthing grid

Gross errors can arise if the evaluation is done based on formula for individual electrodes as shown below.

For vertical rod electrodes,
   ρ = 100ΩM; L = 6M; Φ = 0.05M (2")
   Applying Eqn (1.5), R = 15.5625Ω
   For 18 rods in parallel, R_V = 15.5625/18 = 0.8646Ω

For horizontal electrodes,
  $T = 0.1M$
  N_X = Number of conductors in X direction = 6
  N_Y = Number of conductors in Y direction = 9
  $L_H = 9 \times 50 + 6 \times 80$
     $= 930M$

Applying Ryder's formula (Eqn 1.6), $R_H = 0.2866 \Omega$
  Effective grid resistance $R'_G = R_V \| R_H$
            $= 0.2152 \ \Omega$
But as per Sverak formula {Eqn (1.8)},
  $A = 80 \times 50 = 4000 M^2$
  $L_V = 18 \times 6 = 108M$
  $L = L_H + L_V = 930 + 108 = 1038 \ M$
  $\rho = 100 \Omega M$
  $h = 0.5 \ M$
  $R_G = 0.79 \ \Omega$

It is found that R_G is much greater than R'_G derived from series and parallel combination of horizontal and vertical electrodes. This is due to the fact that resistance areas of earth for individual electrodes are not exclusive and partially overlap which is accounted for in Sverak's formula. If GPR (Ground Potential Rise), Step and Touch potentials are calculated based on R'_G, it could result in unsafe design.

The earth grid resistance, as per Sverak's formula, is not dependent on type of electrode material like Cu, Al or GI. It is also not dependent on cross section of individual electrode. It is a function of physical dimensions (mainly length) and not on physical properties.

1.16 METHODS TO REDUCE EARTH GRID RESISTANCE
From Eqn (1.8), two possibilities exist.
  (i) One method is to reduce soil resistivity (ρ) to a low value. This can be obtained by treating the immediate soil surrounding the electrodes as described in Sec 1.4.
    It must be emphasized that treated earth pits (and retreatment) are required only in case of minor isolated systems like small substations, feeder pillars, lighting poles, HVDS transformers etc. But treating of earth pits may not have much value addition in majority of earthing grids of large Air Insulated Substations (AIS) where the length of buried conductors and area enclosed are very large.

(ii) The second method is to increase the length of buried conductor to the maximum extent possible. Also increase the area enclosed using peripheral conductors (Sec 2.4). This is a one step procedure and the increased cost has to be borne at the beginning.

1.17 MEASUREMENT OF ELECTRODE RESISTANCE

Fall of Potential Method is the most wide spread method used in practice. It consists of two current electrodes (one is electrode under test CET and the other is reference electrode CER) and one potential electrode VE. Refer Figure 1.37.

CET can be a single electrode or earthing grid. Test Current (I) enters here.

CER is a reference electrode at sufficient distance (L) from electrode under test. Test current leaves this electrode.

VE is potential electrode placed between current electrodes. The voltage (V) is measured between CET and VE.

Electrode or Grid Resistance = V/I Ω

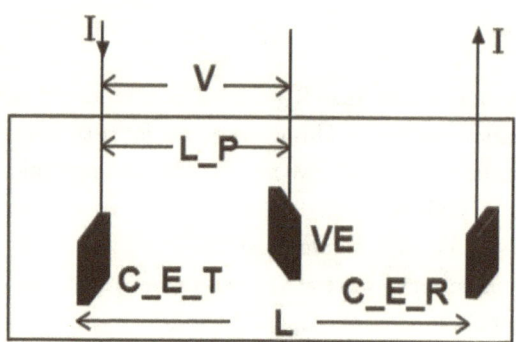

Figure 1.37: Fall of Potential Method

Take three measurements with
   $L\_P = L/2$
   $L\_P = L/2 + D$
   $L\_P = L/2 - D$

If the three readings agree within tolerable accuracy, average of three readings can be taken as electrode or grid resistance.

If the three readings vary widely, the resistance areas described in Sec 1.14.1 interfere with each other. To better the accuracy readings are taken with increased spacing 'L' to minimize interference in resistance areas. Refer Figure 1.38.

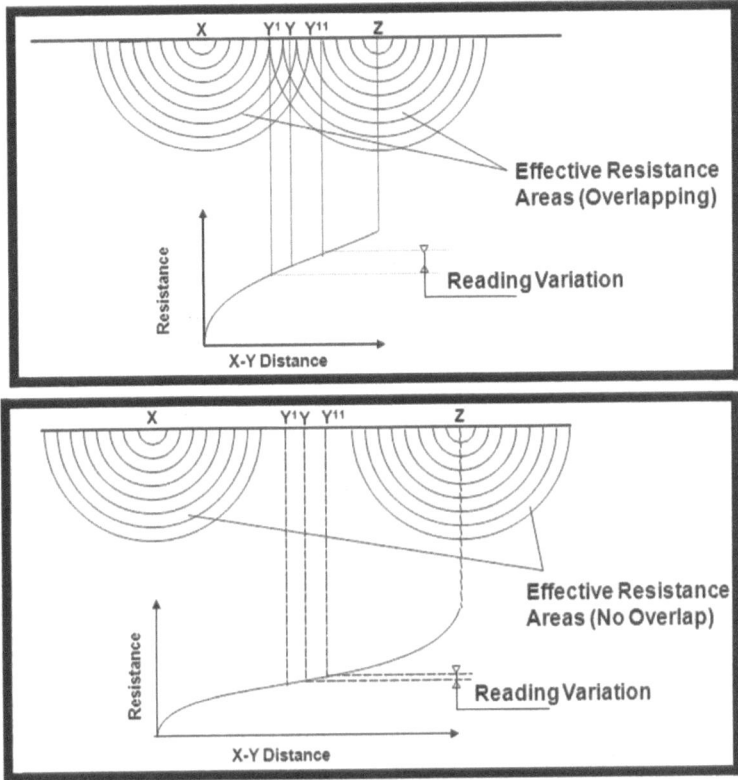

Figure 1.38: Effect of increased distance between current electrodes

In Figure 1.37 all the three electrodes are in the same line. It is not necessary for Potential Electrode VE to be in the same line. In fact better accuracy is obtained in 90° connection in which current leads and potential leads are 90° apart as shown in Figure 1.39.

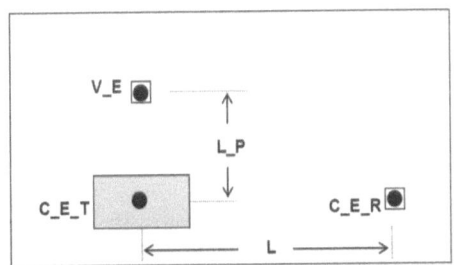

Figure 1.39: 90° Connection

Higher 'L' will yield more faithful results. But there could be limitation due to site constraints. It could vary from a few meters (in small substations) to a few hundred meters in case of large switchyards. D depends on available L and is in the range of 2 to 10 meters. In case of large switch yards where the grid resistance is very small (less than 0.5Ω) it is very difficult to measure grid resistance accurately at site. If switchyard is located in a densely populated urban area, it is hard to locate reference electrode sufficiently away without encroaching outside property. In case of large power plants (like in Figure 2.5), earthing grids of constituent elements like 765kV switchyard, 400kV switchyard, 132kV switchyard, Transformer Yard, TG building, etc are all interconnected in a tight mesh and it is very difficult to isolate one element completely for site measurement purposes. For these reasons, the purported value of earth grid measurement at site suffers from doubtful accuracy.

Hence the measurement of only isolated earth electrodes gives more realistic result. However isolating individual vertical electrodes from grid and measuring its resistance in large switchyard does not have any value addition since effective grid resistance is more influenced by kilometers of horizontal rod or strip electrodes buried in ground. It has become more a ritual to measure annually resistance of individual vertical electrodes. Instead of measuring resistance of every vertical electrode in large switchyard, if at all it can be limited to only those connected to Lightning Arrestors.

1.18 ELECTRODE SIZING

The choices for the type of material and size are only with respect to the amount of fault current to be discharged to earth. The current density (A/mm^2) as per IS-3043 [23] is given below:

Material	Cu	Al	GI
0.5 sec rating	290	178	113
1 sec rating	205	126	80

Table 1.3

The above figures closely match with values given in Cl 7.3, Figure 13 of IEC 60865-1 [32].

Earthing grid for EHV switchyards is designed for 0.5 sec duty and for others 1 sec duty is selected.

As a side remark, we observe that design current density for sizing of copper conductor is typically 3A/mm^2 for *steady state operation* while it is very high for electrode sizing as it is rated only for short time duty (less than a second).

As an example, consider an EHV earthing grid of GI to discharge 40 kA for duration of 0.5 sec.

Current density from Table 1.3 = 113 A/mm²
Section required = 40,000/113 = 353 mm²
Taking corrosion allowance as 15%,
Desired cross section = 353 x 1.15 = 406 mm²
For Strip Electrode, size = 50 x 8 mm; Area – 400 mm²
For Rod Electrode, size =23 mm dia; Area – 415 mm²
The most common sizes used in EHV switchyard are 32 mm and 40 mm dia rods.

If electrode sizing is to be done for any arbitrary time 'T', the corresponding current density is obtained from the following formula:

$$\text{Electrode Rating in A} : \frac{K}{\sqrt{T}} \qquad (1.9)$$

K: Constant defined for 1 sec duty (e.g. 80 for GI)

For example, if GI electrode is to be sized for 0.7 sec,
$I = 80/\sqrt{0.7} = 96$ A/ mm²

Considering mechanical strength and ruggedness requirements, minimum size for main earthing grid shall be 50 mm² for GI and 25 mm² for copper

Except for carrying ground fault current for certain duration, neither the material type nor cross section plays any significant part in earthing grid design. It may be mentioned here that they do not have significant influence on Step and Touch potentials either (Refer Sec 2.12). The choice of material is based on other considerations like possibility of theft, corrosion, etc

Earthing of Electrical System – EHV System

Chapter 2

2.0 INTRODUCTION

In Chapter 1, fundamentals of earthing system like electrode resistance, factors influencing electrode resistance and electrode sizing were discussed. Now we will graduate further towards EHV earth mat design considering human safety [4]. The principles of Ground Potential Rise, Step and Touch potentials are explained. Earthing grid design is explained with a few examples.

2.1 HUMAN ELEMENT

Electric 'shock' is possible only when the human body bridges two points of *unequal* potential. This is the reason why a bird can sit comfortably on a EHV line without getting electrocuted as the voltage between its legs (IZ drop) is insignificant.
Consider a bird sitting on 220kV line.

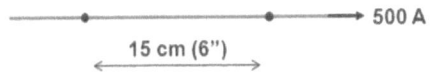

Conductor: Zebra
Impedance: 0.4 Ω/KM
Distance between two feet of bird = 15 cm (6")
Current carried: 500A $\rightarrow \sqrt{3}$ x 220 x 0.5 = 190 MVA
Step voltage across the bird's feet: 500 x 0.4 x 15 x 10^{-5} = 30 mV
This voltage is too small to kill the bird.

If the bird sits on a high voltage object and if its body is close to an earthed point, it will be instantly killed as the voltage across its body will be in kV. Charred body of a crow that got electrocute on 33kV busbar is shown in Figure 2.1.

Figure 2.1: Electrocution of Bird

2.2 HOTLINE OPERATION

This refers to any work carried out on EHV line in live conditions. Typical work carried out are changing insulator strings, washing of insulators, repairing of conductors, repositioning vibration dampers, tightening/replacing jumper bolts, replacement of faulty jumper, etc. There are two methods available – Hot Stick Method and Bare Hand method.

In Hot Stick Method the crew after climbing the tower approaches the line accessories using epoxy hot line sticks and other required tools. The hot line sticks are used to position the conductors appropriately. The crew works from a safe distance from energized line and is at ground potential but near enough to carry out the job. Each hot stick before use must pass a dielectric withstand test of 100kV per foot (30 cm) in accordance to the standards. This method is used up to 400kV level.

In Bare Hand Method, the crew is in physical contact of live wire. The crew wears a conductive hot suit and traverse from ground potential to live potential through insulated bucket or ladder. Insulated platform is normally used in substation outage work. When the platform is raised to the line potential, *there is no potential difference between line and platform*. Refer Figure 2.2. It is similar to bird sitting in the middle of line. The crew can do the maintenance job touching the live wire. While approaching to conductor from ground potential to live potential through insulated bucket or ladder, he shunts the conductor with a metallic clamp (made up of copper) which eliminates the sparking due to ionization of air between the Hot man and live conductor. Hot man is charged at EHV (220 or 400kV) level.

The conducting suit worn by the crew ensures equal potential over the body. The step potential is negligible. With the Hot man wearing conductive hot suit along with hand gloves and hot line shoes the current through body is negligible.

Figure 2.2: Hotline Working

2.3 TOLERABLE CURRENT

Maximum tolerable current for a human body is about 160 mA for one second. If this limit is exceeded, it results in death due to ventricular fibrillation (heart attack). Allowable body current I_B (Amperes), for two body weights, as per IEEE Std-80 [29] is given below:

$$I_B = \frac{0.116}{\sqrt{T_S}} \text{ for a body weight of } 50 \, Kg \qquad (2.1)$$

$$= \frac{0.157}{\sqrt{T_S}} \text{ for a body weight of } 70 \, Kg \qquad (2.2)$$

where T_S is the duration of shock exposure (fault clearance time).
For various exposure times, the withstand currents are as follows:

T_S	I_B (50 Kg)	I_B (70 Kg)
0.2 sec	259 mA	351 mA
0.5 sec	164 mA	222 mA
1.0 sec	116 mA	157 mA

Table 2.1

For shorter duration, body can withstand higher current magnitude. The advantage of modern high speed protection from human safety point of view is evident now as it reduces duration of GPR defined in Sec 2.4. Current and voltage waveforms as captured by a relay during a fault are shown in Figure 2.3. The fault is cleared within 100 msec.

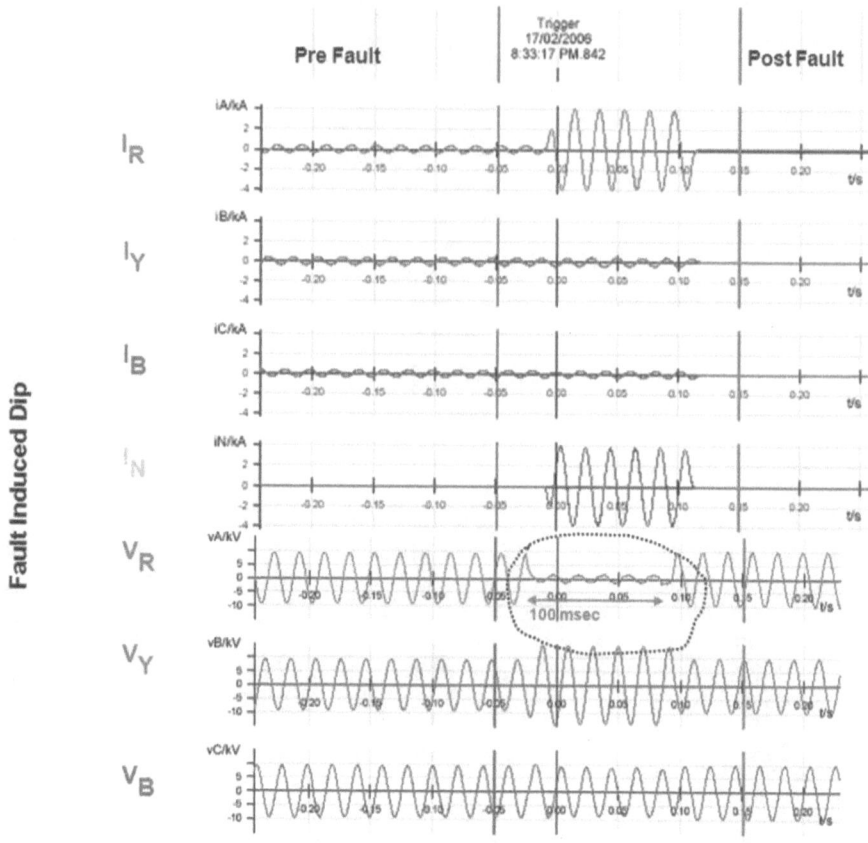

Figure 2.3: Fast Fault Clearance

RCCBs (Residual Current Circuit Breakers) are installed in domestic electrical circuits. The typical setting is 30mA and operating time is almost instantaneous. This offers satisfactory protection with adequate safety margin (30mA against allowable current of more than 100mA).

What kills – voltage or current? Current of certain minimum magnitude and duration results in fatality. Even if the voltage is high, by increasing the resistance between body and earth, the current can be limited below threshold value. The average value of human body resistance (R_B) under dry conditions is 8 to 9KΩ. But for design purposes, conservative value of 1KΩ is assumed, as per IEEE Std-80. Electrical safety shoes and gloves increase the resistance manifold (Mega-Ohms) resulting in very low current through body (Figure 2.4). Also insulating mats are provided in front of electrical panels so that even if a live part is accidentally touched the current through the body is very low.

The mats made of elastomer (rubber, latex, etc) are 3 to 4 mm thick and offers resistance in hundreds of Mega-Ohms even when wet (Figure 2.5).

Figure 2.4: Person Working with Safety shoes & Hand gloves

Figure 2.5: Insulating rubber mat in front of Electrical Panels

It is pertinent here to describe the operation of ubiquitous 'tester' used by electrician. It has a neon tube with high resistor in series. When the top of tester is pressed with finger and the other end is connected to live 240V supply we are establishing a current path through our body to ground. That current is very small but enough to light the neon lamp but within safe value for humans. The glowing of neon lamp indicates the test point is live. Tester shall be used with bare hands. With insulated gloves or shoes the current may be too small for neon lamp to glow even if the test point is live.

2.4 GROUND POTENTIAL RISE

Ground Potential Rise (GPR) is the voltage to which the earth mat is going to rise when it discharges the current. I_G is the current discharged to earth and R_G is the earth grid resistance (Figure 2.6),

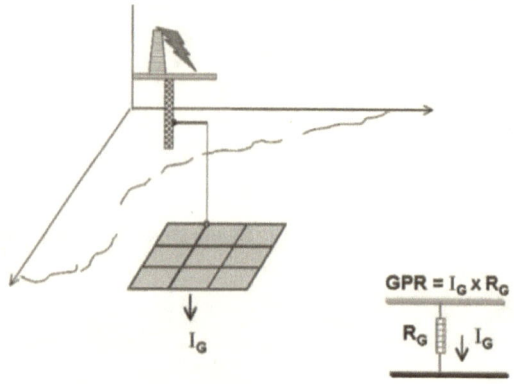

Figure 2.6: Ground potential Rise

$$GPR = I_G R_G \qquad (2.3)$$
$$I_G = K I_F \qquad (2.4)$$

K is called Split Factor.

I_F is the fault current. *Only that fraction of fault current that gets discharged to earth (I_G) contributes to GPR.* A very low I_G results in low GPR. This is the basis for the dictum – "Get the fault current back to source neutral via metal". In this case, K is nearly zero and I_G is insignificant. Typical examples are in oil platforms, ships and EHV underground cables. If GPR is very low (nearly zero), touch and step potentials are irrelevant.

Another parameter to lower GPR (Eqn 2.3) is to achieve very low R_G through Earthing Grid design. For example, in Figure 2.7 outline of earthing grid provided for a large power plant site is shown. Concentrated Energy Centre areas like EHV Switchyard, Transformer yard, TG Building, etc with their own earthing grid are shown as rectangular boxes.

With the peripheral conductors enclosing a very large area, R_G is significantly low (<0.2Ω). Refer Sec 1.16 for influence of Area on earth grid resistance.

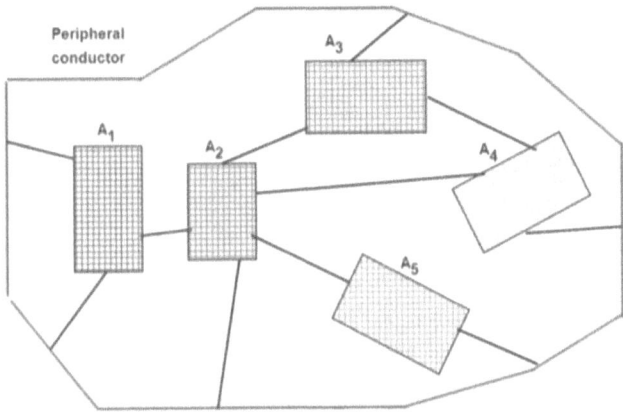

Figure 2.7: Earthing Grid for a large power plant

2.5 CURRENT DISCHARGED TO EARTH - I_G

To emphasize the point that I_G is always *not* equal to ground fault current (I_F), three cases are considered.

2.5.1 Case 1

Fault is within the switchyard and transformer connection is star–delta (Figure 2.8). It is very unusual to have delta on EHV side but given here only to illustrate a point. Entire fault current is discharged to earth to return to source 2. In this case, K = 1.

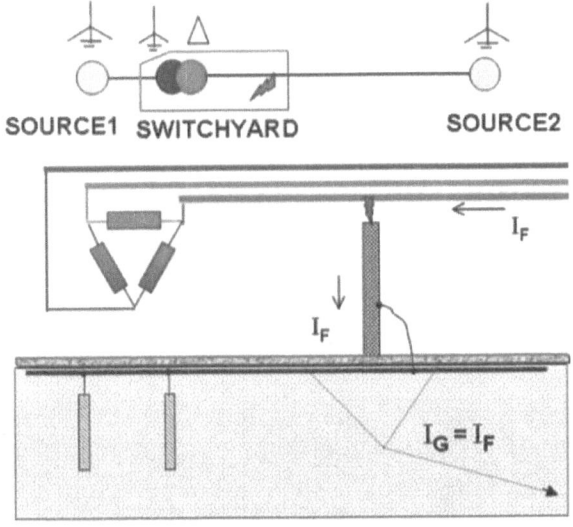

Figure 2.8: Fault within Substation (Star-Delta)

2.5.2 Case2

Fault is within the switchyard and transformer connection is delta–star (Figure 2.9). Part of the fault current (I_{F1}) returns to local transformer via metallic conductor (earth mat) and does not contribute to GPR. The other part (I_{F2}) is discharged to earth to return to source 2 and contributes to GPR in remote station. In this case, K < 1.

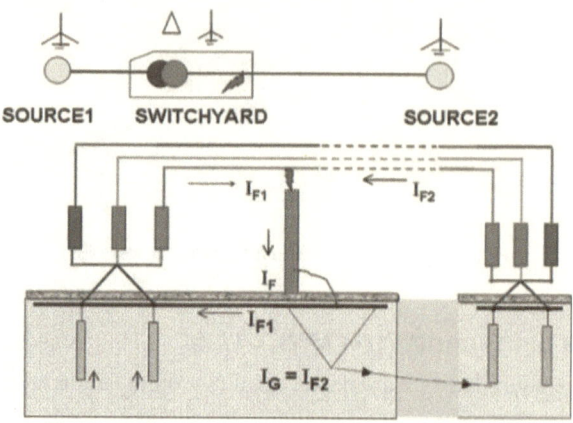

Figure 2.9: Fault within Substation (Delta-Star)

2.5.3 Case 3

Fault is on transmission line and transformer connection is delta–star (Figure 2.10). Part of the fault current (I_{F1}) returns to transformer at source 1 via earth and contributes to GPR. The other part (I_{F2}) returns to source 2 via earth and contributes to GPR at the other switchyard. In this case also, K < 1.

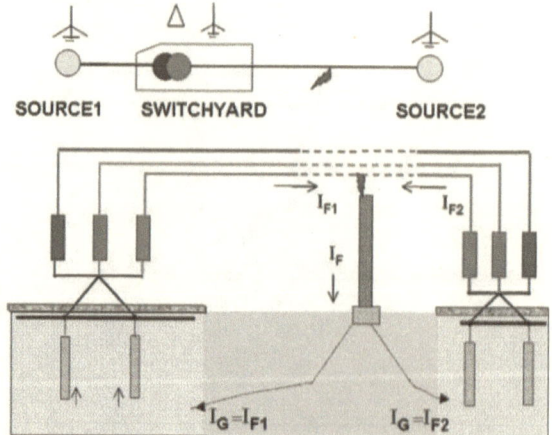

Figure 2.10: Fault outside Substation (Delta-Star)

2.5.4 Application of I_{FAULT} and I_{GRID}

I_{FAULT} is used primarily for conductor (electrode) sizing calculations. Refer Sec 1.18. I_{GRID} is for used for evaluating step and touch voltages in earthing grid calculations. If I_{FAULT} is used in earthing grid calculations, it will result in over sized design of earthing grid.

2.6 FAULT ON TRANSMISSION LINE

In Sec 2.5.3, we ignored the effect of earth wires (or shield wires). If earth wire is considered (Figure 2.11), part of ground fault current returns to source via metallic earth wires. The other part that returns via earth *only* contributes to GPR. It shall be ensured that the earth wire on end tower shall have a *direct metallic connection* to switchyard earth mat (last mile connectivity).

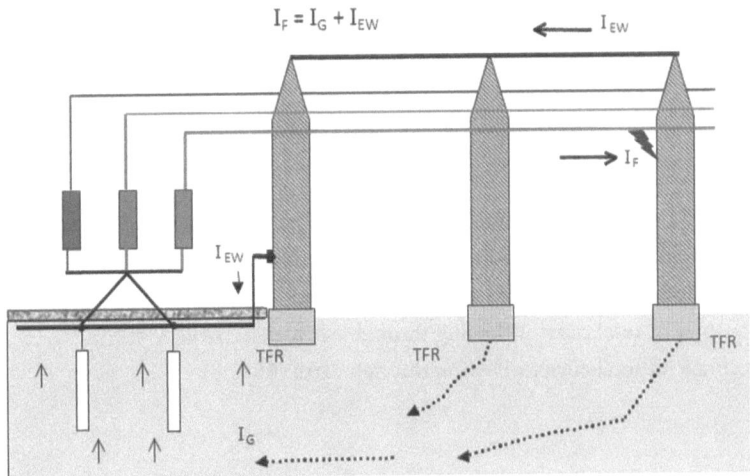

Figure 2.11: Transmission line with earth Wire

Current diverted to earth wire (I_{EW}) depends on:
  (i) Number of earth wires: 1 for 220kV and below; 2 for 400kV and above
  (ii) Material of earth wire (GI, Al)
  (iii) Typical size of earth wire:
      (a) 7/3.15 GI for 220kV and below
      (b) 7/3.66 GI for 400kV and above.
      (c) Where OPGW is used, 12/7 AACSR,
  (iv) Tower footing resistance (5 to15Ω)

Current diversion to earth wire happens in two ways: due to induction and conduction.

2.6.1 Current Diverted by Induction

Refer Figure 2.12. When fault current flows on phase conductor, due to mutual induction current is induced in earthwire. Current Diverted to Earth Wire by Induction can be calculated using software that calculates line constants [20].

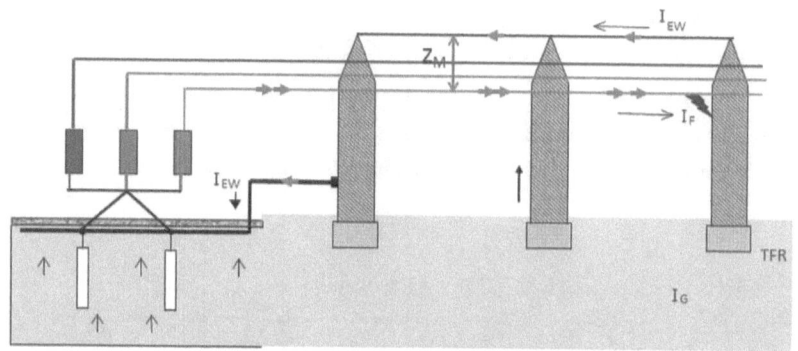

Figure 2.12: Fault current diverted by induction

In Figure 2.13, 220kV double circuit configuration with one earth wire is shown. Phase conductor is zebra and earth wire is dog. Dimensions are in Meters. Using software, current diverted by induction is obtained as follows:

Percentage of total current flowing through earth wire: 32%
Percentage of total current flowing through earth: 68%

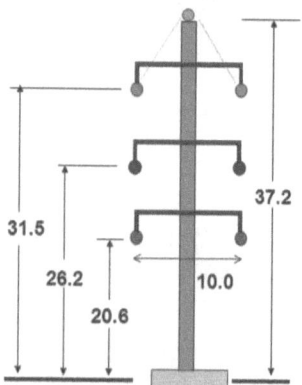

Figure 2.13: 220kV D/C Configuration with one earth wire

In Figure 2.14, 400kV twin moose single circuit configuration with two earth wires (EBB3/8) is shown. Dimensions are in Meters. Bundle Radius is 45 cm. Using software, current diverted by induction is obtained as follows:

Percentage of total current flowing through first earth wire: 26%
Percentage of total current flowing through second earth wire: 26%
Percentage of total current flowing through earth: 48%
Refer Appendix 2-1 for theory.

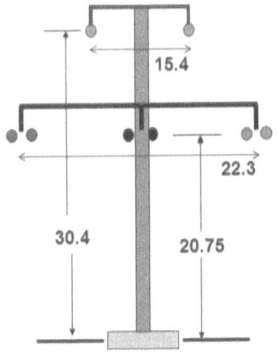

Figure 2.14: 400kV S/C Configuration with two earth wires

2.6.2 Current Diverted By Conduction

Refer Figure 2.15. On the faulted tower, current has two paths to return – one through over head earth wire and the other (IG_1) to earth through the tower footing resistance. The current IG_1 returning towards feeding substation encounters next tower. Here also two paths are available – one up the tower to earth wire and the other again through earth (IG_2). This process continues till the substation is reached. Finally part of fault current returns by conduction through earth wire and the rest returns through earth.

The solution to this problem is obtained by solving ladder network.

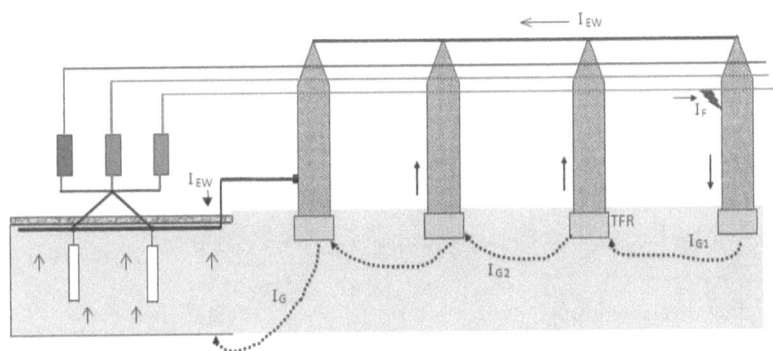

Figure 2.15: Fault current diverted by conduction

If length of ruling span is less, more number of towers is available for the current to return via metallic earth wire and correspondingly the proportion of current returning via earth is less. Conversely, If span length is more, less number of towers is available for the current to return via metallic earth wire and correspondingly the proportion of current returning via earth is more. Refer Table 2.2.

Span (M)	100	200	300	400
I_G (%)	68	76	80	82

Table 2.2

The legs of transmission tower act like vertical ground electrodes. Some times additional auxiliary electrodes or counterpoises (horizontal strip electrodes) are also connected to tower legs. Counterpoise is a horizontal electrode of GI, 15m to 25m long buried 750 mm below ground and connected to every tower leg. The effective resistance offered by tower leg to earth is termed as Tower Footing Resistance (TFR). Typically TFR is between 5Ω to 15Ω.

One advantage of having low TFR is to reduce the probability of 'back flashover' during direct lightning stroke to tower. This will be covered later in future Chapter on Insulation Coordination.

Another advantage is that it offers low resistance path for the fault current to reach the earth wire on the top at each tower location, thus increasing the proportion of current returning via earth wire and correspondingly the proportion of current returning via earth is less. Refer Table 2.3.

Span (M)	200	200	300	300
TFR (W)	10	5	10	5
I_G (%)	76	69	80	74

Table 2.3

If many transmission lines converge in a switchyard, current diverted through conduction further reduces.

2.6.3 Current Discharged to Earth for ESTEP & ETOUCH

I_F: Fault Current

I_G = Current discharged to earth for calculation E_{STEP} & E_{TOUCH}

$$I_G = K_1 \times K_2 \times I_F \tag{2.5}$$

K_1 = Current diverted through induction

K_2 = Current diverted through conduction

As a first order approximation, we can assume $K_1 = 0.7$ and $K_2 = 0.8$.
$I_G = 0.7 \times 0.8 \times I_F \cong 0.6 \times I_F$

To verify that fault current actually returns through earth wire, CBCT is mounted on earth wire on 220kV end tower as shown in Figure 2.16. In field, mounting of CBCT is not an easy task as it has to be supported and mounted at a height of 50M. The CBCT is connected to an over current relay. For an external fault in system, part of fault current returns through earth wire and the current as recorded in relay is shown in Figure 2.17.

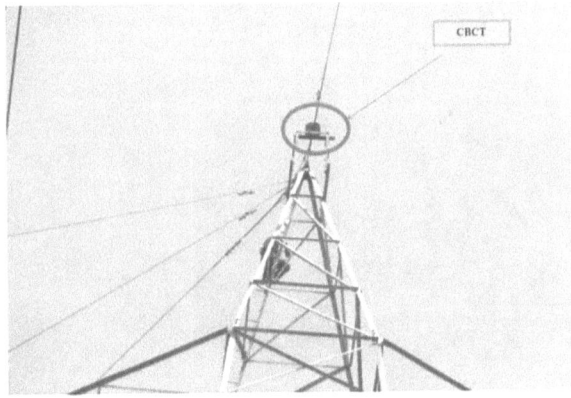

Figure 2.16: CBCT on Earth Wire

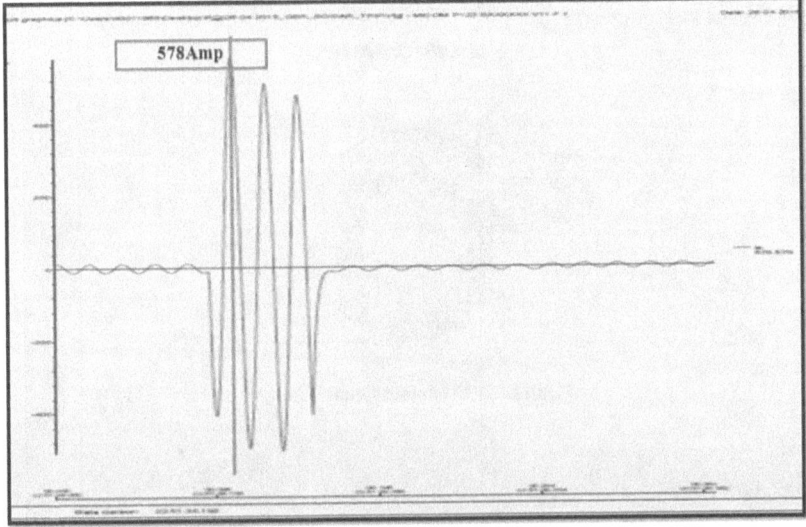

Figure 2.17: Current recorded in Relay

2.7 FAULT ON EHV CABLE

Typical cross section of EHV cable is shown in Figure 2.18. It consists mainly of inner copper conductor core surrounded by XLPE insulation with lead or Aluminum sheath over the insulation. The sheath offers mechanical protection to XLPE insulation against external intrusion. In addition, it offers metallic path for return of fault current back to source. Its function is same as armour used in LV and MV cables.

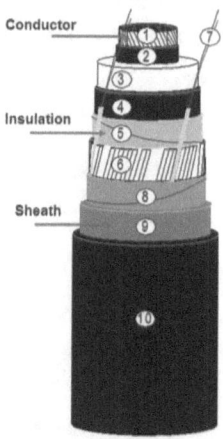

Figure 2.18: EHV Cable

Typical cross sectional view of Corrugated Aluminum sheath of 220kV 1200 mm² Cu cable is shown in Figure 2.19.

Sheath C/S Area

Figure 2.19: Sheath Cross Section

d_1 = 107 mm
d_2 = 107 − 2 x 2.1 = 102.8 mm
A = $(\pi/4)$ $(107^2 − 102.8^2)$
  = 692 Sq MM

Assume fault currrent I_F = 40 KA
Current density = 40,000/692 = 58 A/mm²

As per IS 3043 (Table 1.3), allowable current density for 0.5 sec rating is 178 A/mm². This is much higher than requirement of 58 A/mm², thus ensuring adequate safety margin.

Figure 2.20 shows a cable with cross bonding of sheath at every one third location in a section. Details of EHV cable and associated cable engineering including cross bonding will be covered in future chapter on cables. In LB1 (Link Box 1 of first minor section) sheath of R phase is connected to Y phase, Y phase to B phase and B phase to R phase. In LB2 (Link Box 2 of second minor section) another sheath transposition is done. At LB3 (Link Box3, end of major section), all the sheaths are earthed to ground.

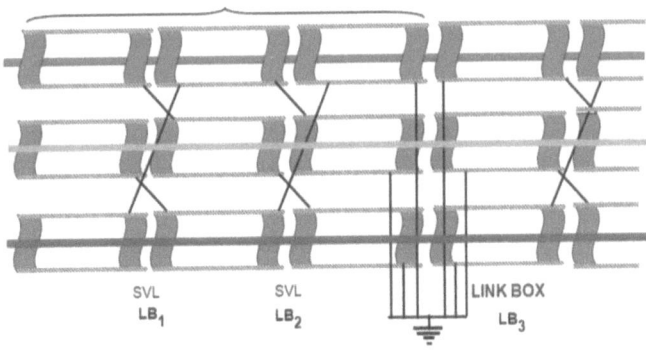

Figure 2.20: Sheath Cross Bonding

The distribution of fault current returning to source is illustrated in Figure 2.21. The current distribution in the main conductor as well sheath is shown. It can be seen that only upto the end of first major section from fault (LB3), the fault current returns through sheath of one phase. Beyond LB3, sheaths of all three phases are available for fault current to return. For the cable shown in Figure 2.19, from the faulted point till LB3 the fault current returns through Aluminum Sheath of 692 mm² cross section and beyond LB3 the available cross section is 2076 mm² (3 x 692).

At the end of first major section at LB3, the fault current has two alternate paths to return, one via the earth and the other through Aluminum sheath with equivalent cross section of 2076 mm². The resistance offered by metallic path is so low that very little current discharges to earth and practically the entire current returns via metal (Aluminum sheath). In this case K is very small. For earthing grid design purposes, usually split factor of 10% is assumed for calculating Step and Touch potentials.

I_G = 0.1 x I_F

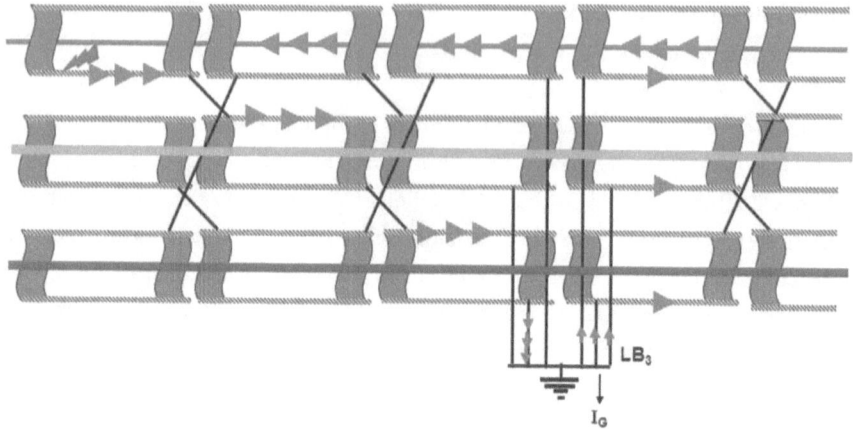

Figure 2.21: Fault current distributions in sheath

To verify that fault current actually returns through sheath (like earth earthwire in O/H line), three CBCTs are mounted on three sheaths connected to ground at the end of cable section (similar to LB3).

Refer Figure 2.22. The CBCTs are connected to over current relays. For a distant fault in system (not in cable itself), part of fault current returns through Aluminum sheath. The registered current wave form is shown in Figure 2.23. Current through each sheath is almost 1kA peak.

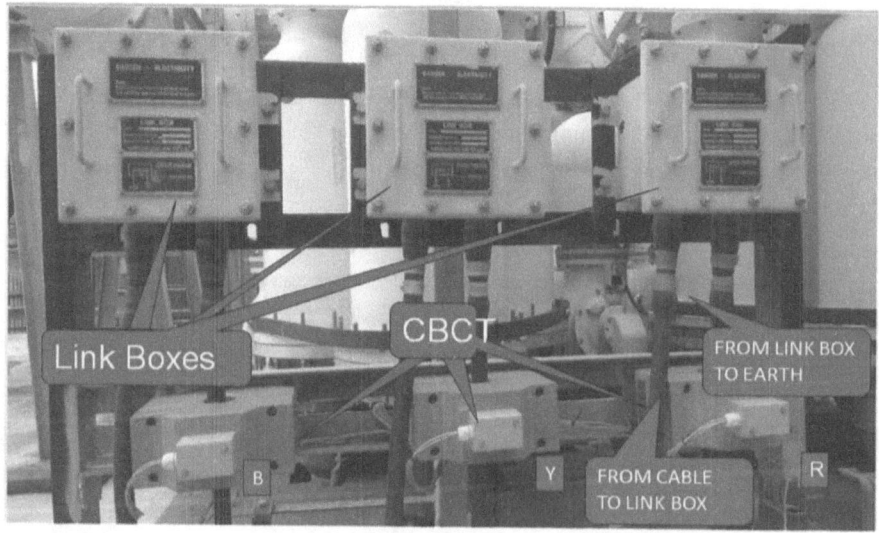

Figure 2.22: CBCT mounted on Link Box.

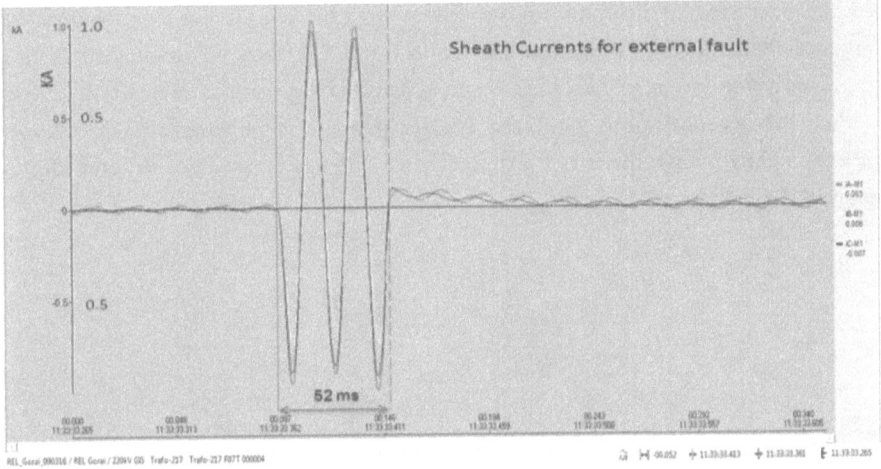

Figure 2.23: Registered fault current in sheath

2.8 STEP AND TOUCH POTENTIALS

Step and Touch potentials refer to the potential experienced by a person standing on the surface when earth mat buried, say 750 mm, below surface has risen to GPR. All the points of earthing grid formed by metallic structure are almost at the same potential (GPR). However the surface potential will vary at every point depending on resistivity of native soil, surface layer resistivity (gravel) if present and earthing grid geometry. Step potential is the difference in surface potentials experienced by person bridging a distance of 1 meter with his feet without contacting any other grounded object. Touch potential is the difference between GPR and the surface potential at the point where person is standing, while his hand is in contact with grounded structure. Refer Figure 2.24.

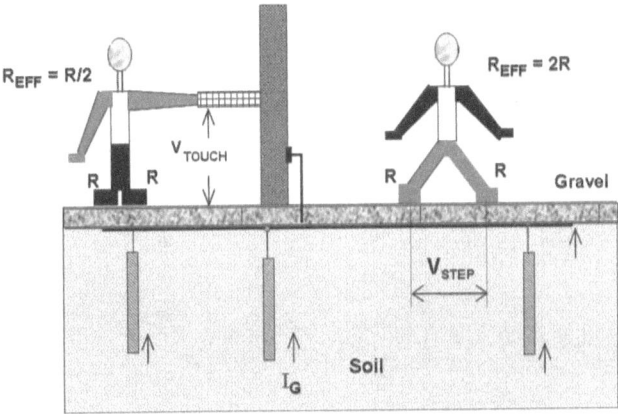

Figure 2.24: Step and Touch potentials

If resistance offered by each foot is R, intuitively it can be seen that for step potential the resistance offered is 2R while for touch potential it is R/2. Hence the deciding criteria for design will be touch potential as less resistance is involved. *Step potential is usually academic.*

Also another subtle difference is that the step potential is the difference between two surface potentials while the touch potential is the difference between GPR and surface potential. Refer Figure 2.25.

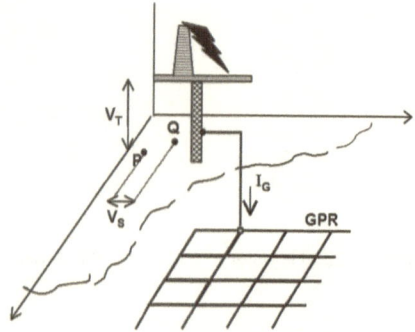

Figure 2.25: GPR, Step and Touch potentials

2.8.1 Step Potential

In Figure 2.26, surface potential as percentage of GPR is plotted for the profile line shown in earthing grid. Step potential is the difference between potentials at P and Q, 1 meter apart.

It can be seen that maximum step potential is at the corners where one foot is within boundary and the other is outside boundary. In this case there is a steep change in surface potential.

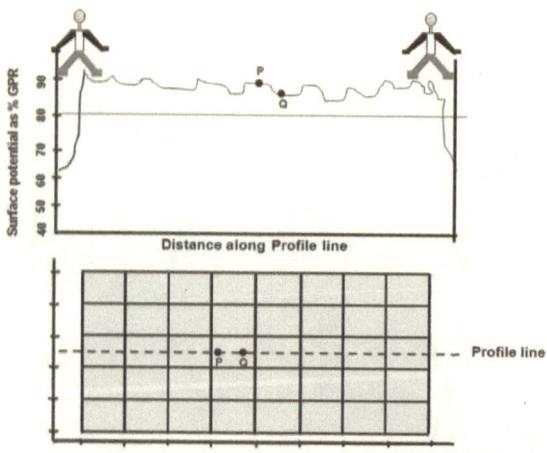

Figure 2.26: Surface Potential along profile line

We will compare performance of two grids shown in Figure 2.27 – 6x9 grid and the other 2x2 grid. Earth grid resistance (R_G) can be worked out as per Sverak formula (Eqn 1.8) and found to be almost same in both cases. In the later case though length of buried conductor is less (340M vs 1038M), area enclosed is much higher (7200M² vs 4000M²).

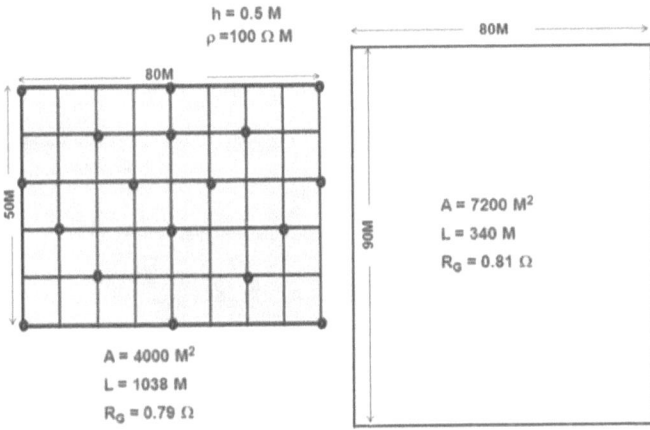

Figure 2.27: Two grids with same RG

In Figure 2.28, surface potential for the profile line is shown for 6x9 grid.

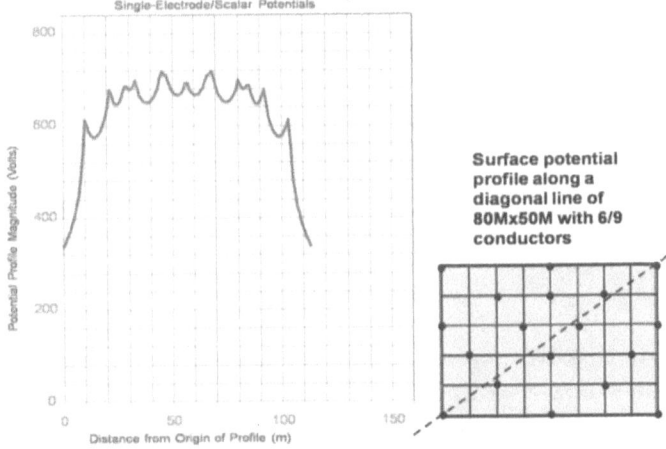

Figure 2.28: Surface potential for the profile line for 6x9 grid

In Figure 2.29, the derived step voltage for the profile line is shown. The maximum step voltage is 64V near the edge of grid.

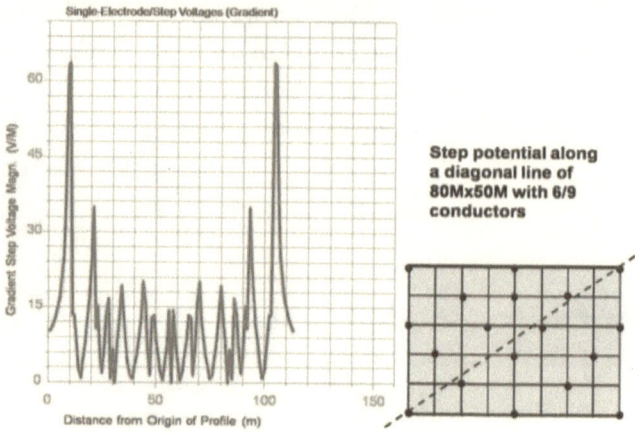

Figure 2.29: Step Voltage for 6x9 grid

In Figure 2.30, surface potential for the profile line is shown for 2x2 grid.

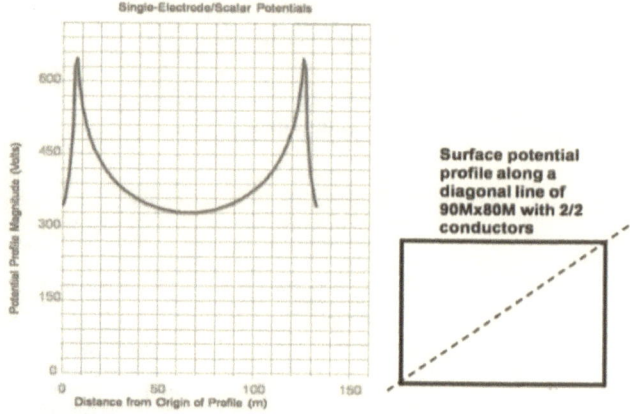

Figure 2.30: Surface potential for the profile line for 2x2 grid

In Figure 2.31, the derived step voltage for the profile line is shown. The maximum step voltage is 120V near the edge of grid. This is almost twice that obtained for 6x9 grid. The advantage of forming earthing grid with a number of horizontal electrodes in X and Y direction is evident now.

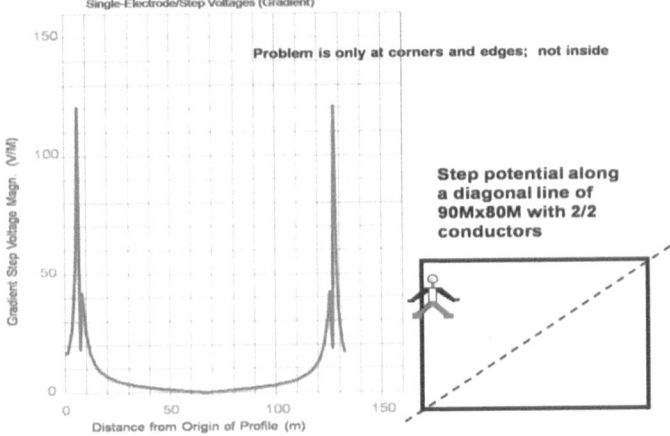

Figure 2.31: Step Voltage for 2x2 grid

2.8.2 Touch Potential

Touch potential is the difference between GPR and the surface potential. GPR, surface potential and touch potential are shown in Figure 2.32 for 6x9 grid. In the middle of profile line (60M) touch voltage is only about 50V.

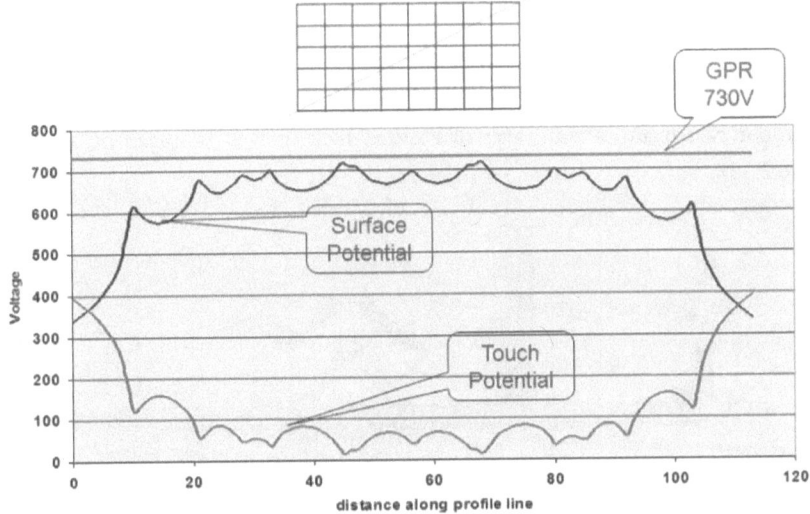

Figure 2.32: GPR, Step amd Touch Voltage for 6x9 grid

GPR, surface potential and touch potential are shown in Figure 2.33 for 2x2 grid. In the middle of profile line (60M) touch voltage is almost 500V which is ten times more than that in 6x9 grid. The advantage of forming earthing grid with a number of horizontal electrodes in X and Y direction is strikingly clear now.

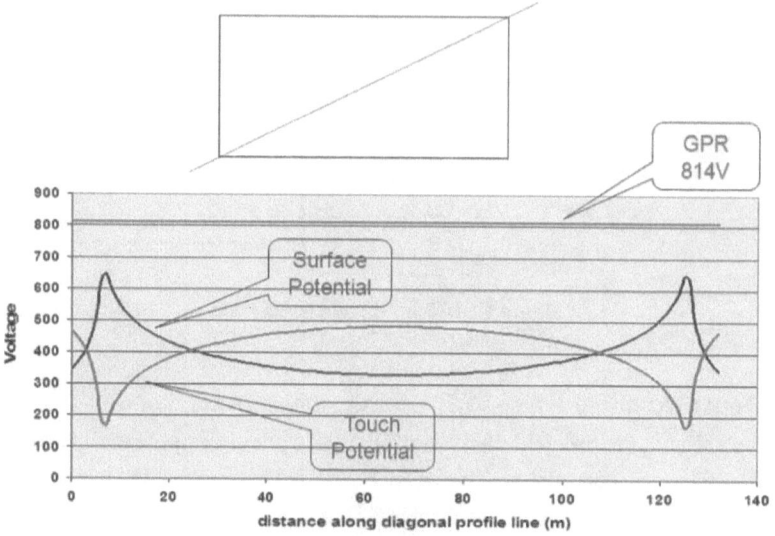

Figure 2.33: GPR, Step amd Touch Voltage for 2x2 grid

2.9 EFFECT OF THIN LAYER OF CRUSHED ROCK

In outdoor switchyard, a thin layer of crushed rock (gravel) is spread on the surface (Figure 2.34).

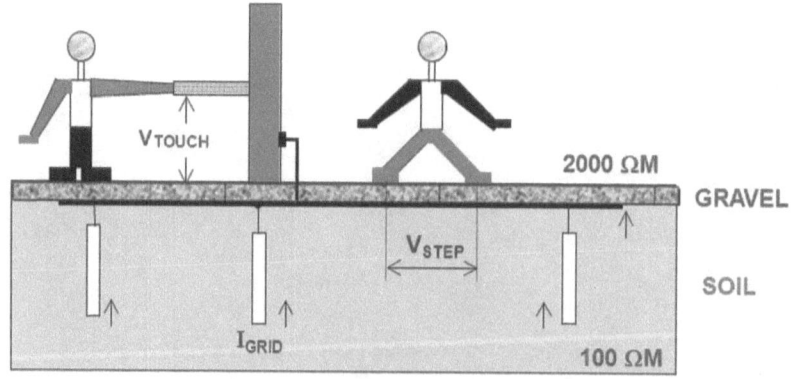

Gravel is not mandatory in all switchyards

Figure 2.34: Switchyard paved with Gravel

Resistivity of gravel (ρ) is about 2000ΩM while that of native soil is about 100ΩM. Since ρ of gravel is high, only a high voltage can force the current through the body to cause injuries. The gravel acts like insulator and throws the electric field generated by GPR back to the soil. The Reduction Factor (C_S) defines the percentage of electric field that reaches the surface. It is used in calculation of allowable step and touch potentials.

C_S – Reduction Factor
$C_S = f(h,K)$
h = Thickness of surface layer of gravel (Typically 3" to 6" = 8 to 15 CM)
K = Reflection Factor

$$K = \frac{\rho - \rho_{SL}}{\rho + \rho_{SL}} \tag{2.6}$$

ρ = Native Soil Resistivity
ρ_{SL} = Surface Layer (Gravel) Resistivity

Reflection Factor with different combinations of ρ and ρ_{SL} are given in Table 2.4. For example, if surface layer (gravel) resistivity is 2000Ω-M and soil is 100Ω-M, K is 0.9 implying only 10% electric field reaches the surface (Figure 2.35).

Gravel	Reflection Factor K = (ρ - ρ_{SL})/(ρ + ρ_{SL})				
ρ_{SL} (ΩM)	ρ: 50 ΩM	ρ: 100 ΩM	ρ: 150 ΩM	ρ: 200 ΩM	ρ: 300 ΩM
1000	-0.826	-0.818	-0.739	-0.667	-0.539
1500	-0.936	-0.875	-0.818	-0.765	-0.667
2000	-0.951	-0.905	-0.879	-0.818	-0.739
2500	-0.961	-0.923	-0.887	-0.852	-0.786
3000	-0.967	-0.935	-0.905	-0.875	-0.818

Table 2.4

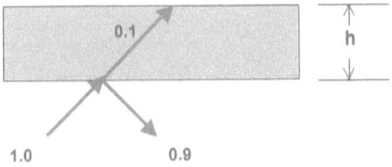

Figure 2.35: Reflection factor due to gravel

It is possible to work out reduction factor C_S as a function of Reflection Factor and surface layer thickness as given in Figure 11 of IEEE Std-80, 2013

But a more straight forward method is to use following formula:

$$C_S = 1 - \frac{0.09\left(1 - \frac{\rho}{\rho_{SL}}\right)}{2h + 0.09} \tag{2.7}$$

If no gravel is spread, $\rho_{SL} = \rho_{SOIL}$, $C_S = 1$. Gravel is not *mandatory*. If I_G or R_G is low, GPR itself will be low and there will not be need for gravel to achieve safe Step and Touch potentials,

2.10 ALLOWABLE TOUCH AND STEP POTENTIALS

The allowable touch and step potentials are given by following formula:

$$E_{STEP}^{LMT} = (R_B + 6 \times \rho_{SL} \times C_S) I_B \tag{2.8}$$
$$E_{TOUCH}^{LMT} = (R_B + 1.5 \times \rho_{SL} \times C_S) I_B \tag{2.9}$$

R_B: Body resistance, assumed as 1000Ω
ρ_{SL}: Surface layer resistivity, $2000\Omega M$ (typical for gravel)
I_B: Permissible body current (Table 2.1)
C_S: Reduction factor (0 to 1) from Eqn (2.7)

2.10.1 Example

Find E_{STEP} and E_{TOUCH} given following data:.

(i) Weight of man = 70 KG
(ii) Fault duration = 0.5 sec
(iii) Soil Resistivity $\rho = 100\Omega M$
(iv) Surface layer (Gravel) Resistivity $\rho_{SL} = 2000\Omega M$
(v) Gravel Thickness h = 0.1m (4")/0.05m (2")

Solution:.

From Eqn (2.7),
Reduction factor:
$C_S = 0.56$ for h = 4"
$C_S = 0.26$ for h = 2"
From Table 2.1,
Allowable body current I_B: 0.222A
Body Resistance as per IEEE Std-80
$R_B = 1000\Omega$

Allowable Step and Touch Potentials, from Eqn(2.8) and Eqn(2.9), are given in Table 2.5.

Surface Layer Thickness (h)	Allowable Step Potential E_{STEP}^{LMT}	Allowable Touch Potential E_{TOUCH}^{LMT}
4"	2109V	694V
2"	1697V	591V

Table 2.5: Allowable Step and Touch potentials

As stated earlier, Touch Potential is deciding criterion. If it is satisfied, Step Potential is automatically satisfied as $E_{STEP}^{LMT} > E_{TOUCH}^{LMT}$

Also with reduction in gravel thickness, allowable step and touch potentials go down. 10 to 15 years after switchyard is commissioned the thickness of gravel may have reduced due to natural attrition. It is preferable to resurface with fresh layer of gravel so that allowable step and touch potentials are maintained as per original design values.

In Eqns (2.8) and (2.9), factors 1.5 and 6 are used. They are derived based on finding the resistance of electrode representing a foot. Refer Figure 2.36. Each foot is represented an electrode of solid disk of 8 cm diameter. The electrode resistance of the disk is given by:

$R_F = \rho/4D = \rho/4 \times 0.08 \cong 3\rho$

For step potential, both feet resistance come in series, hence the factor is 6 (3+3).

For touch potential, both feet resistance come in parallel, hence the factor is 1.5 (3/2).

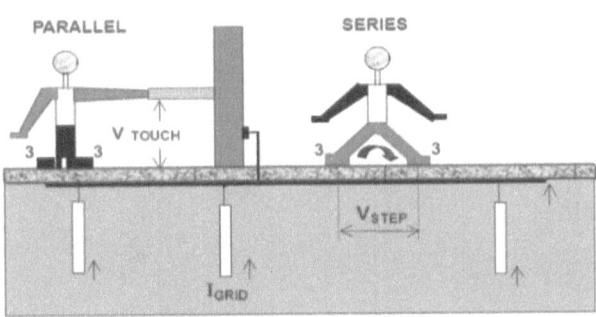

Figure 2.36: Resistance of human foot

The earthing grid is designed such that design touch and step potentials are less than the limits obtained from Eqns (2.8) and (2.9).

2.11 DESIGN TOUCH AND STEP POTENTIALS

For the given grid geometry, the following factors are evaluated (as per IEEE Std 80):

$$K_1 = \frac{D^2}{16hd}; \; K_2 = \frac{(D+2h)^2}{8Dd}$$

$$K_3 = \frac{h}{4d}; \; K_4 = \frac{1}{\sqrt{(1+h)}}$$

$$K_5 = \frac{8}{[\pi(2n-1)]}; \; K_6 = \frac{1}{2h}$$

$$K_7 = \frac{1}{(D+h)}; \; K_8 = \frac{(1-0.5^{N-2})}{D}$$

Geometrical factor for Step Voltage

$$K_s = \frac{1}{\pi}(K_6 + K_7 + K_8)$$

Geometrical factor for Touch Voltage

$$K_T = \left(\frac{1}{2\pi}\right)[ln(K_1 + K_2 - K_3) + K_4 ln(K_5)]$$

Irregularity factor:

$$K_I = 0.644 + 0.148N$$

Design step voltage:

$$E_{STEP}^{DESN} = \frac{(\rho I_G K_T K_I)}{(L_S)} \tag{2.10}$$

Design touch voltage:

$$E_{TOUCH}^{DESN} = \frac{(\rho I_G K_T K_I)}{(L_M)} \tag{2.11}$$

D = Spacing between parallel grid conductors, M
d = Diameter of grid conductor, M
h = Depth of grounding grid, M
N = Number of equivalent conductors
I_G = Current discharged into the earth, A
L_X = Maximum length of horizontal conductor in X direction, M
L_Y = Maximum length of horizontal conductor in Y direction, M
L_H = Total length of horizontal conductors, M
L_{VI} = Length of individual vertical rod, M

L_V = Total length of vertical rods, M
L_P = Total length of peripheral conductors, M

$$L_M = L_H + \left[1.55 + 1.22 \left(\frac{L_{VI}}{\sqrt{L_X^2 + L_Y^2}}\right)\right] L_V$$

$L_S = 0.75 L_H + 0.85 L_V$
N = Number of equivalent conductors
  = $2 L_H/L_P$
ρ = Soil Resistivity, ΩM

The design is acceptable if $E_{STEP}{}^{DSN} < E_{STEP}{}^{LMT}$ and $E_{TOUCH}{}^{DSN} < E_{TOUCH}{}^{LMT}$.
In the above formula, step and touch potentials are evaluated at the corner of grid where the values are maximum. Using software, it is possible to obtain the step and touch potentials at any point on the grid as illustrated in Figure 2.28 to Figure 2.33.

2.11.1 Example

To meet touch and step potential limits worked out in Sec 2.10.1 following grid parameters were assumed:
  I_G = 15kA; h = 0.75M
  d = 37.5 mm
    = 0.0375M

Grid pattern is as shown in Figure 2.37. Only Typical conductors are shown.

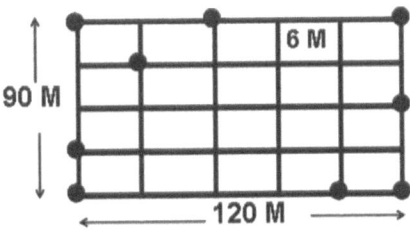

Figure 2.37: Selected Grid Pattern

Spacing: D = 6M
Vertical electrodes (•): 18 Nos, 6M long
  L_V = 6M
  L_V = 6 x 18
    = 108M

Number of conductors in X direction

$$N_X = \left(\frac{90}{6}\right) + 1 = 16$$

Number of conductors in Y direction

$$N_Y = \left(\frac{120}{6}\right) + 1 = 21$$

Total length of Horizontal conductors
L_H = (21 x 90) + (16 x 120)
 = 3810M

L_P = Total length of peripheral conductors, M
L_P = 2(90+120) = 420M
N = Number of equivalent conductors
 = (2 x 3810)/420 = 18.1429

Approximate value of N can also be found following equation:

$N \cong \sqrt{N_X N_Y} = 18.3$

K_1 = 80.2139; K_2 = 31.3336

K_3 = 5.0134; K_4 = 0.7559

K_5 = 0.0722; K_6 = 0.6667

K_7 = 0.1481; K_8 = 0.1667

K_I = 3.3291

K_S = 0.3124

K_T = 0.4267

Finally,

$E_{STEP}{}^{DSN}$ = 529V

$E_{TOUCH}{}^{DSN}$ = 535V

For 2" gravel as surface layer, allowable limits are

$E_{STEP}{}^{LMT}$ = 1697V

$E_{TOUCH}{}^{LMT}$ = 591V

Selected design is safe since $E_{STEP}{}^{DSN} < E_{STEP}{}^{LMT}$ and $E_{TOUCH}{}^{DSN} < E_{TOUCH}{}^{LMT}$. As expected, the deciding criterion is Touch Voltage.

Ground Grid Resistance and GPR

From Sverak formula, Eqn (1.7) & (1.8)

$L = 3810+108 = 3918M$

$A = 90 \times 120 = 10800M^2$

$h = 0.75M$

$R_G = 0.45\Omega$

$GPR = I_G \times R_G = 15000 \times 0.45 = 6750V$

Summary of results (2" gravel)

GPR	Step Potential		Touch Potential	
	E_{STEP}^{LMT}	E_{STEP}^{DSN}	E_{TOUCH}^{LMT}	E_{TOUCH}^{DSN}
6750	1697V	529	591V	535

2.12 INFLUENCE OF CROSS SECTION

From Eqn (2.10), it is seen that *step potential is independent of diameter (cross section)*. K_6, K_7 and K_8 and hence K_S do not involve diameter (d).

The variation of touch potential with diameter, using Eqn (2.11), is shown in Figure 2.38. For 400% increase in diameter, the reduction in touch potential is only 40%.

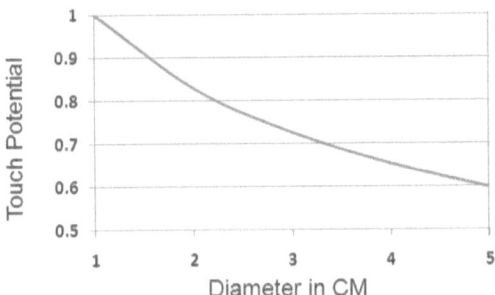

Figure 2.38: Touch Potential vs Conductor Size

Thus it is concluded that cross section has minor influence on Step and Touch potentials while the linear dimension (length) has significant impact.

Cross section plays a part only when sizing the electrode to carry the specified fault current for a given duration (Sec 1.18). Otherwise the influence of cross section is marginal on earthing grid design.

2.13 SUMMARY OF SWITCH YARD GROUND GRID DESIGN

1. From plot plan, estimate area available for grounding
2. Measure soil resistivity before starting any plant activity. Refer Sec 1.6.
3. Evaluate fault level at the given location (system studies) to obtain I_{FAULT}
4. Evaluate split factor (S_F) to get current discharged to ground I_{GRID}. Refer Sec 2.6.3 and Sec 2.7. $I_{GRID} = S_F \times I_{FAULT}$
5. Size the conductor based on I_{FAULT} and t_{FAULT} (0.5 or 1 sec). Refer Sec 1.18
6. Make preliminary grid design using horizontal and vertical rods and surface layer (e.g. gravel) of given thickness.
7. Find allowable body current (I_B) for given t_{FAULT}. Refer Sec 2.3.
8. Calculate allowable limits for Step and Touch potentials E_{STEP}^{LMT} and E_{TOUCH}^{LMT}. Refer Sec 2.10.
9. Based on assumed grid geometry find design Step and Touch potentials E_{STEP}^{DSN} and E_{TOUCH}^{DSN}. Refer Sec 2.11.
10. If $E_{STEP}^{DSN} < E_{STEP}^{LMT}$ and $E_{TOUCH}^{DSN} < E_{TOUCH}^{LMT}$ design is complete. Otherwise, redesign till design voltages are within safe limits.
11. If design is inadequate, modify the grounding grid by increasing horizontal grid length (L_H) and or vertical rod length (L_V) by increasing the number of vertical electrodes. Refer Figure 2.39.

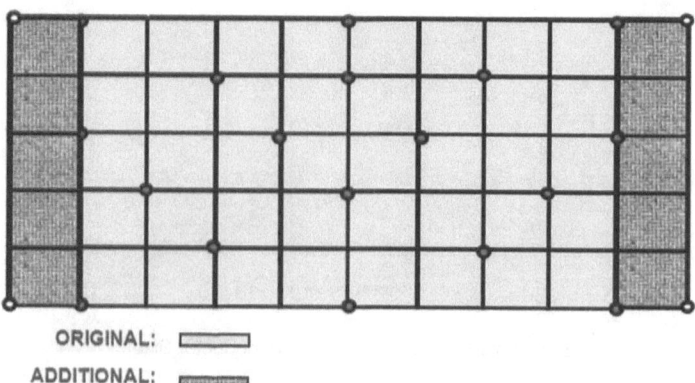

Figure 2.39: Grounding Grid Design

In Figure 2.37 symmetrical grid is assumed (spacing between horizontal electrodes in X direction and Y direction is uniform). Because of simplicity in design computation this has been widely used in the past. But with availability of software (like CDEGS, ETAP,

Autogrid, etc) with powerful computational features with easy GUI, it is not mandatory to design with symmetrical grid pattern.

For example, it is more effective to increase the number of conductors near the periphery with less conductors in the middle as shown in Figure 2.40. Intuitively this makes sense as the problems of touch and step potentials are more acute near the corners and edge of grid rather than in middle of grid.

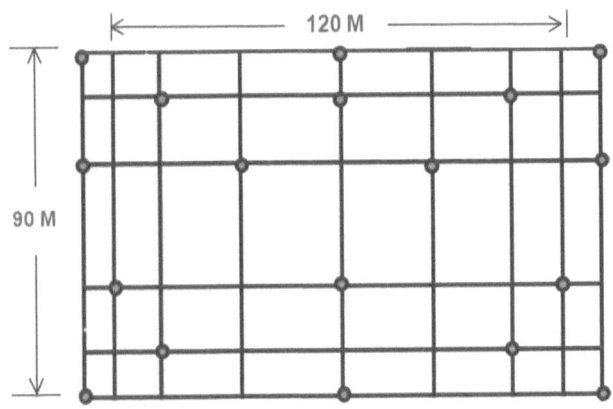

Figure 2.40: Asymmetrical Grounding Grid

2.14 EARTHING OF SPECIAL AREAS

2.14.1 Isolator/Disconnector Operation platform

In EHV switchyard only equipment that needs to be locally operated is isolator or disconnector. During the manual operation, there could be a fault in equipment due to any of the following reasons:
- Opening an energized circuit
- Mechanical failure or jam
- Electrical breakdown of insulator of isolator
- Electrical breakdown when interrupting line charging current or transformer no load (magnetizing) current

During the above occurrence the operator should not experience step and touch potentials that are not safe. If the earthing grid is well designed step and touch potentials in all areas within the substation are already within limits. But as a measure of abundant caution, an auxiliary earth mat is created just below the MOM (Motor Operated Mechanism) of isolator where the operator will stand. All the parts of isolator to be earthed are directly connected to the main earthing grid as well as auxiliary earth mat which in turn is again connected to main earthing grid of substation.

Since the auxiliary earth mat on which operator stands is metallically connected to main earthing grid there is no potential difference between the two resulting in negligible touch potential. Typical auxiliary earth mat connection is shown in Figure 2.41 and 2.42.

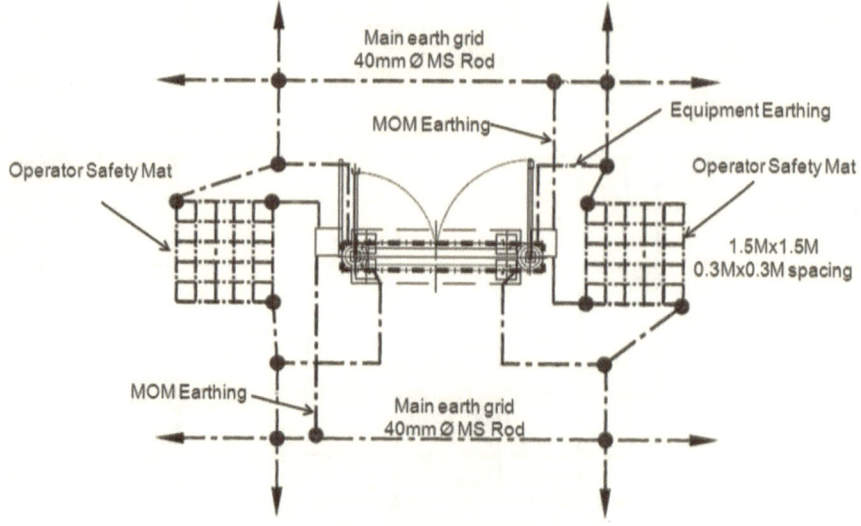

Figure 2.41: Isolator Earthing - Plan

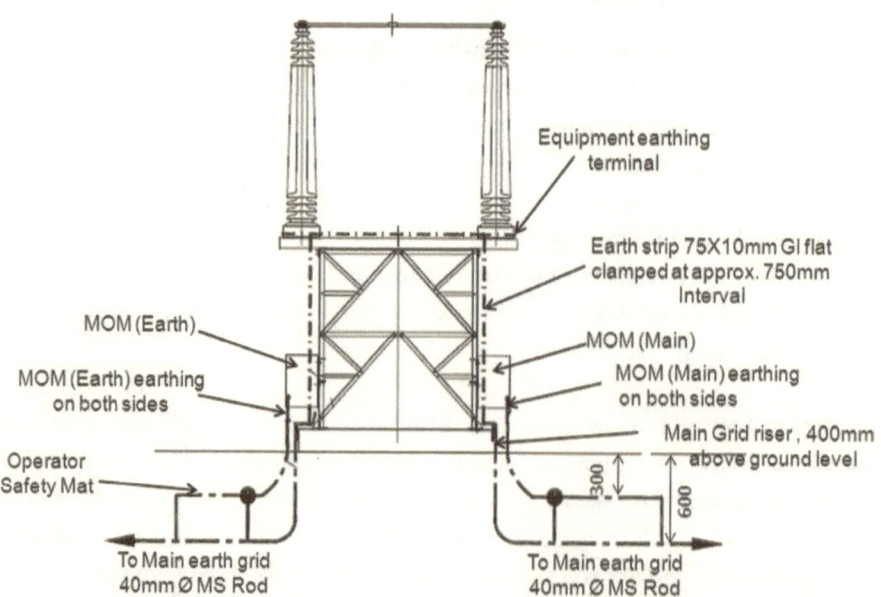

Figure 2.42: Isolator Earthing – Elevation

2.14.2 Fence Earthing

The fence provided for switchyard should be within grounding grid area. The outer most earthing conductor shall be about 1 meter away from fence. The gravel shall also extend up to fence. The fence shall be connected to main earthing grid at every support structure or at periodic interval of say every 20 meters. Below EHV line crossing if fence is present, the fence shall be earthed at that location.

In large power plants this is not an issue as the EHV switchyard is well within the plant boundary and peripheral earthing conductors extend well beyond switchyard fence. Refer Figure 2.43.

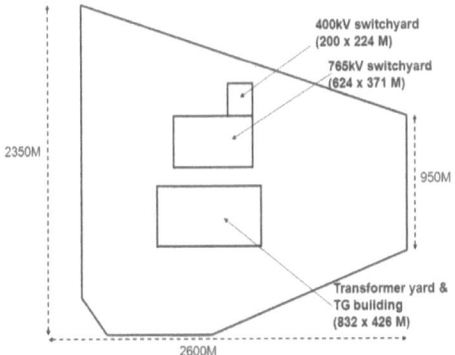

Figure 2.43: Large Power Plant Plot Plan

2.15 SAMPLE OUTPUTS FROM SOFTWARE (CDEGS)

2.15.1 Earthing Studies for Ultra Mega Power Plant (UMPP)

Results of case studies done for earthing grid design are shown below. UMPP is spread over a wide area covering approximately 3kM x 3kM. The main Energy Centre areas are 765kV & 400kV switchyard, Transformer Yard and TG Building. Refer Figure 2.44

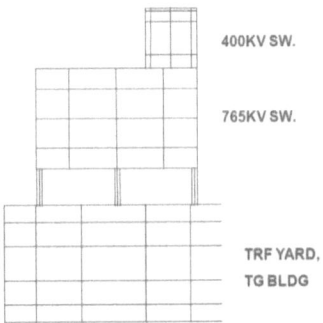

Figure 2.44: Large Power Plant Energy Centre Areas

If earthing is grid designed considering only Energy Centre areas, some areas outside Energy Centre areas are unsafe from Touch Potential point of view. Refer Figure 2.45.

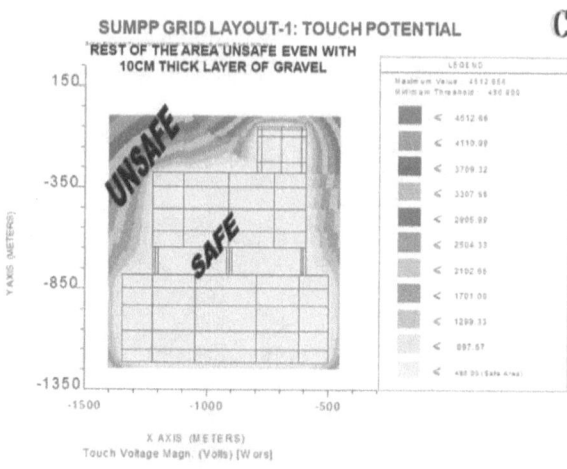

Figure 2.45: Earthing grid only in Energy Centre Areas

The studies were repeated with peripheral conductors outside Energy Centre areas as shown in Figure 2.46

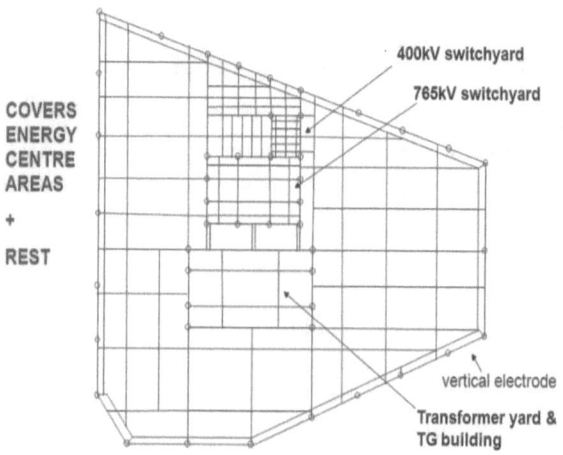

Figure 2.46: Earthing grid with peripheral conductors

Entire plant area is safe in this case as shown in Figure 2.47. In this case within switchyard touch and step potentials are within limits even without gravel in switchyard but as a measure of abundant caution gravel is spread in switchyard and transformer yard. Since the area enclosed is very high due to presence of peripheral conductors, earth grid resistance (R_G) is very low resulting in low GPR.

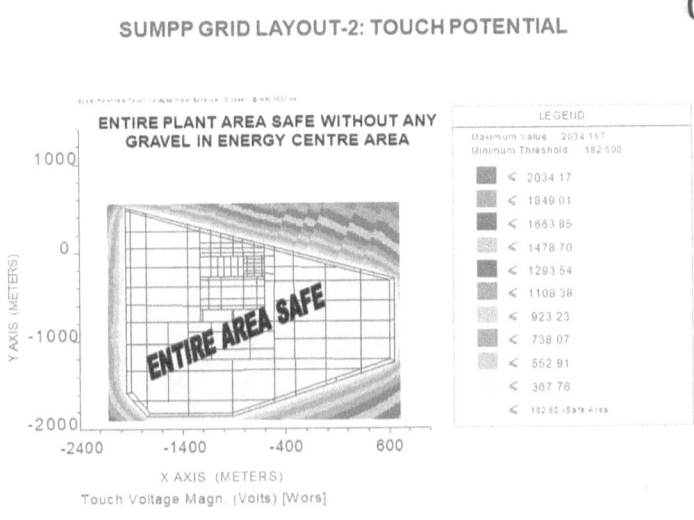

Figure 2.47: Earthing grid with peripheral conductors

2.15.2 Raw Water Intake Pump House

We have encountered the system already. Refer Figure 1.11. Based on soil resistivity measurements, three layer model for soil is obtained from the software. The soil resistivity ρ is 238ΩM between surface and 1.5M below ground, 37ΩM between 1.5M and 3.7M below ground and 224ΩM below 3.7M. It is advantageous to bury the earthing grid just below 1.5M as resistivity (ρ) is lowest. Refer Figure 2.48.

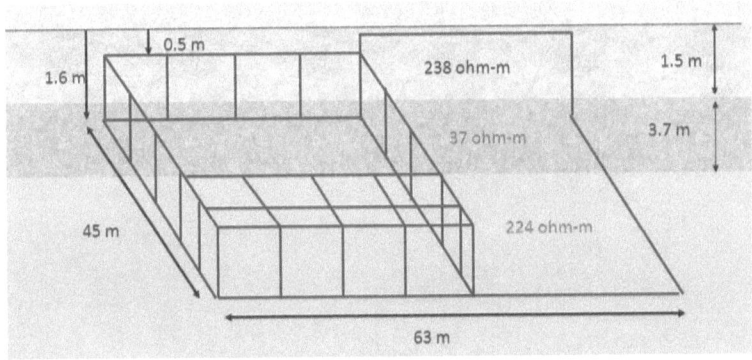

Figure 2.48: Earthing grid at depth of 1.5M

In Figure 2.49, touch potential profile is given when earthing grid is buried 0.8M below ground. Majority of areas are unsafe from touch potential point of view.

Figure 2.49: Earthing grid at depth of 0.8m

The same case was repeated when earthing grid is buried 1.6M below ground. Refer Figure 2.50. Entire area is safe. This is achieved by cleverly using low resistivity soil stratum for laying the earthing grid. Of course the excavation effort in this case will be more.

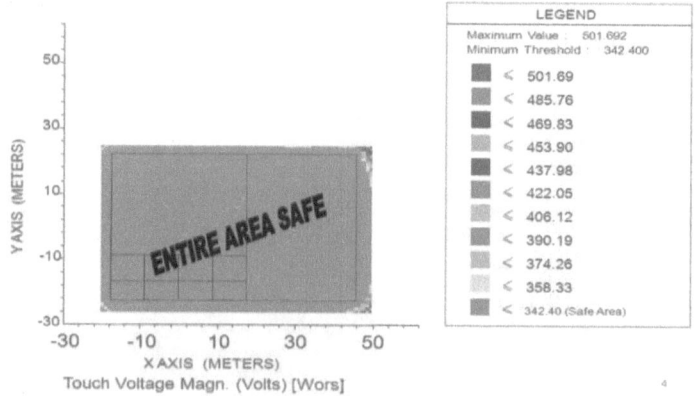

Figure 2.50: Earthing grid at depth of 1.6m

2.16 GIS EARTHING

Typical earthing layout of indoor 220kV GIS is shown in Figure 2.51. In this figure 220kV cable cellar is located below ground floor (basement) and GIS is located on ground floor. In some layouts, 220kV cable cellar is located on ground floor and GIS is located on first floor. Top floors accommodate Control and Relay panels and MV cable cellar, MV switchgears and associated panels.

GIS earthing comprises two distinct parts – one that is in touch with soil below ground and the other is below finished floors above ground.

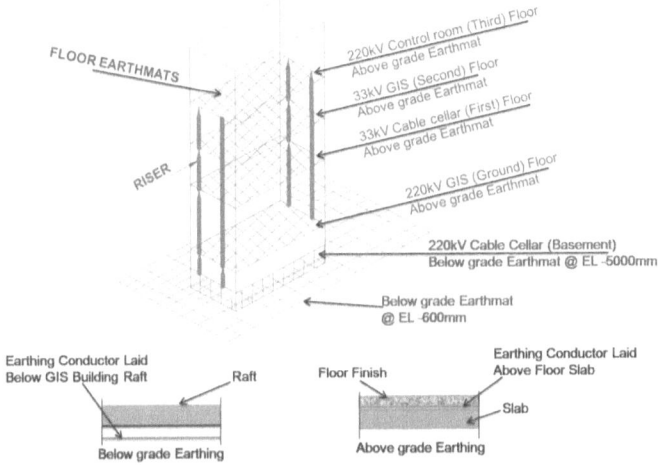

Figure 2.51: Indoor GIS layout

2.16.1 Below Ground Earthing

Grids in touch with soil are usually placed in two locations – 600 mm to 750 mm below ground level and below the raft which could be 3 to 6 metres below ground. The two are interconnected by vertical conductors. Refer Figure 2.52.

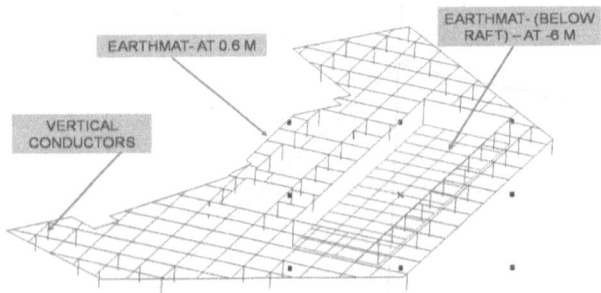

Figure 2.52: Earth Mat below ground

The design principles of the AIS grounding (estimation of step and touch potentials) are applicable here also. Since GIS occupies less than 10% of AIS for the same number of bays, area available for grid as well as length of horizontal conductors that can be buried are not large. A number of long vertical electrodes may be required to achieve desired resistance (R_G). GIS is installed mostly in densely populated areas. In these GIS substations the external connections are through EHV cables. As seen in Sec 2.7, with cable connection, split factor is less than 10% that results in very low current discharged to earth (I_G).

Ground Potential Rise (GPR) = $I_G \times R_G$

Since I_G is low, GPR is not excessive even if R_G is high.

Risers from below ground earthing reach to all the floors above. GIS enclosures are connected to the risers at frequent intervals. GIS bays are connected to earth mat using GI flats.

2.16.2 Above Ground Earthing

On the floor housing GIS and other floors above, GI mesh of small spacing (typically 1M) is embedded below FFL (Finished Floor Level). GI is preferred as it is compatible with rebars in concrete. No adverse effect has been observed in practice if earth mesh below FFL and rebars are bonded together. This is termed as 'Equipotential grounding grid'. This is connected to the risers coming from below the ground earthing grid.

GIS enclosures of all three phases are bonded together. Also enclosures are earthed at frequent intervals (as per manufacturer's recommendation). Multiple earthing and reduced mesh spacing reduce the intensity of VHFT (Very High Frequency Transient) induced voltages. VHFT (in GHz range) are of extremely short duration and *are not harmful to humans*. VHFT in GIS are generated due to switching operation of disconnector or breaker.

2.16.3 Earthing connection arrangement between GIS/Transformer and EHV Cable Termination

The GIS enclosure under normal conditions carry current almost equal to current in main phase conductor. The situation is similar to case in IPBD (Isolated Phase Bus Duct) in large power plants. In IPBD, enclosure currents even under normal operating condition can be of the order of 10kA to 20kA. However in case of GIS, current magnitudes are much less (usually less than 2kA) since operating voltage is much higher. To prevent GIS enclosure current entering cable termination, insulating barrier is provided between GIS enclosure and Cable termination.

Since the barrier introduces electrical discontinuity, sometimes sparks are observed due to VHFT caused by switching operation or through faults. To suppress the sparks, special earthing arrangements are adopted [15].

(a) EHV cable connecting GIS at one substation to GIS another substation (Refer Figure 2.53A). The cable sheaths are connected to earth at both ends. The insulating barrier is bridged by Surge arrestor which conducts if voltage across the insulator increases beyond set value and prevents sparking. The site installation is shown in Figure 2.53B. A few manufacturers of cable termination do not insist on provision of varistor as the barrier thickness is too high for the arc to spark over.

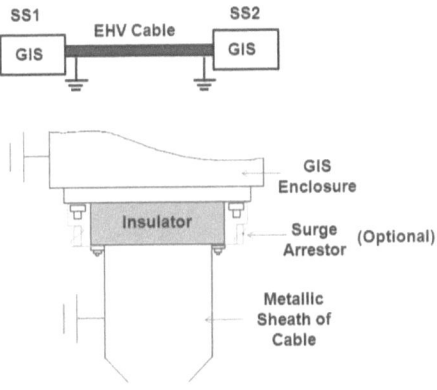

Figure 2.53 A: Earthing arrangement for GIS to cable

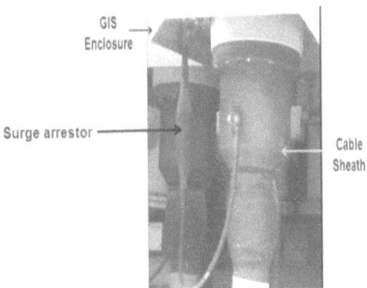

Figure 2.53 B: Site Installation Photo

(b) EHV cable connecting GIS at a substation to Transformer at the same location (Figure 2.54A). The earthing arrangement at GIS end is same as in Figure 2.53A. But at transformer end, cable sheath is connected to earth via SVL (Surge Voltage Limiter). Under normal conditions SVL will not conduct and the cable sheath is earthed only at one end (GIS end). The insulating barrier is bridged by Surge arrestor which conducts if voltage across the insulator increases beyond set value and prevents sparking. The site installation is shown in Figure 2.54B. Only the connections at higher level (say 6M) elevation are shown. No surge arrestor is provided across insulating barrier. SVL is connected between sheath and earth bus near ground level in Link Box. A few manufacturers of cable termination do not insist on provision of varistor as the barrier thickness is too high for the arc to spark over.

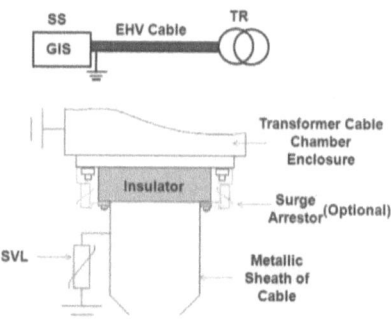

Figure 2.54 A: Earthing arrangement for EHV cable to Transformer

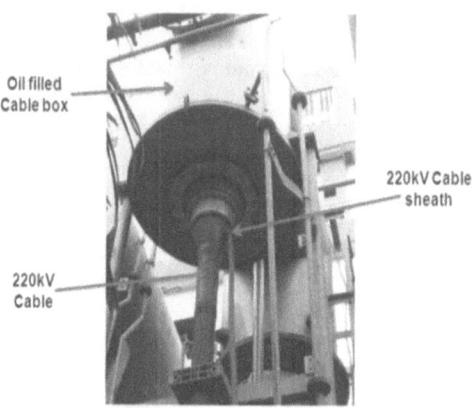

Figure 2.54 B: Site Installation Photo

Division of Current Between Ground Wire and Earth

Appendix 2-1

2-1.1. INDUCTION

2-1.1.1 One Ground Wire

$$\begin{bmatrix} V_a \\ V_b \\ V_c \\ 0 \end{bmatrix} = \begin{bmatrix} Z_{aa} & Z_{ab} & Z_{ac} & Z_{ag} \\ Z_{ba} & Z_{bb} & Z_{bc} & Z_{bg} \\ Z_{ca} & Z_{cb} & Z_{cc} & Z_{cg} \\ Z_{ga} & Z_{gb} & Z_{gc} & Z_{gg} \end{bmatrix} \begin{bmatrix} I_a \\ I_b \\ I_c \\ I_g \end{bmatrix} \quad (2\text{-}1.1)$$

Let $I_a = I_b = I_c = I_0$
$3I_0 + I_g + I_d = 0$

$(2\text{-}1.2)$

I_g: Current in ground wire
I_d: Current through earth

From (2-1.1)
$I_o (Z_{ga} + Z_{gb} + Z_{gc}) + I_g Z_{gg} = 0$

$$I_g = -I_0 \left[\frac{Z_{ga} + Z_{gb} + Z_{gc}}{Z_{gg}} \right]$$

$(2\text{-}1.3)$

From (2-1.2) & (2-1.3),
$I_d = -3I_o - I_g$

$$= -I_0 \left[3 - \frac{Z_{ga} + Z_{gb} + Z_{gc}}{Z_{gg}} \right]$$

Let $k_1 = \dfrac{Z_{ga} + Z_{gb} + Z_{gc}}{Z_{gg}}$

$(2\text{-}1.4)$

From (2-1.3) & (2-1.4)

$$\left|\frac{I_g}{I_d}\right| = \left|\frac{k_1}{3-k_1}\right| = k_2$$

$I_g = k_2 I_d$

$I_g + I_d = 1.0$

$I_d = 1.0 - k_2 I_d$

$I_d = \dfrac{1}{1+k_2}$: Fraction of total current through earth.

$I_g = 1 - I_d$: Fraction of total current through ground wire.

2-1.1.2 Two Ground Wires

$$\begin{bmatrix} V_a \\ V_b \\ V_c \\ 0 \\ 0 \end{bmatrix} = \begin{bmatrix} Z_{aa} & Z_{ab} & Z_{ac} & Z_{ag} & Z_{aw} \\ Z_{ba} & Z_{bb} & Z_{bc} & Z_{bg} & Z_{bw} \\ Z_{ca} & Z_{cb} & Z_{cc} & Z_{cg} & Z_{cw} \\ \hdashline Z_{ga} & Z_{gb} & Z_{gc} & Z_{gg} & Z_{gw} \\ Z_{wa} & Z_{wb} & Z_{wc} & Z_{wg} & Z_{ww} \end{bmatrix} \begin{bmatrix} I_a \\ I_b \\ I_c \\ I_g \\ I_w \end{bmatrix}$$

(2-1.5)

Let $I_a = I_b = I_c = I_0$

From eqn (2-1.5), last two rows,

$$[0] = [Z_1]\begin{bmatrix} I_0 \\ I_0 \\ I_0 \end{bmatrix} + [Z_2]\begin{bmatrix} I_g \\ I_w \end{bmatrix}$$

$$[Z_2]\begin{bmatrix} I_g \\ I_w \end{bmatrix} = -[Z_1]\begin{bmatrix} I_0 \\ I_0 \\ I_0 \end{bmatrix}$$

$$\begin{bmatrix} I_g \\ I_w \end{bmatrix} = -[Z_2]^{-1}[Z_1]\begin{bmatrix} I_0 \\ I_0 \\ I_0 \end{bmatrix}$$

(2-1.6)

$$Z_2 = \begin{bmatrix} Z_{gg} & Z_{gw} \\ Z_{wg} & Z_{ww} \end{bmatrix}$$

$$[Z_2]^{-1} = \frac{1}{Z_{gg}Z_{ww} - Z_{wg}Z_{gw}} \begin{bmatrix} Z_{ww} & -Z_{gw} \\ -Z_{wg} & Z_{gg} \end{bmatrix}$$

$$= \begin{bmatrix} Y_{gg} & Y_{gw} \\ Y_{wg} & Y_{ww} \end{bmatrix}$$

$$[Z_2]^{-1}[Z_1] = \begin{bmatrix} Y_{gg} & Y_{gw} \\ Y_{wg} & Y_{ww} \end{bmatrix} \begin{bmatrix} Z_{ga} & Z_{gb} & Z_{gc} \\ Z_{wa} & Z_{wb} & Z_{wc} \end{bmatrix}$$

$$= \begin{bmatrix} (Y_{gg}Z_{ga} + Y_{gw}Z_{wa}) & (Y_{gg}Z_{gb} + Y_{gw}Z_{wb}) & (Y_{gg}Z_{gc} + Y_{gw}Z_{wc}) \\ (Y_{wg}Z_{ga} + Y_{ww}Z_{wa}) & (Y_{wg}Z_{gb} + Y_{ww}Z_{wb}) & (Y_{wg}Z_{gc} + Y_{ww}Z_{wc}) \end{bmatrix}$$

(2-1.7)

From eqns. (2-1.6) & (2-1.7)

$$\begin{bmatrix} I_g \\ I_w \end{bmatrix} = -I_o \begin{bmatrix} Y_{gg}\sum_{i=a}^{c} Z_{gi} + Y_{gw}\sum_{i=a}^{c} Z_{wi} \\ Y_{wg}\sum_{i=a}^{c} Z_{gi} + Y_{ww}\sum_{i=a}^{c} Z_{wi} \end{bmatrix}$$

$$3I_o + I_g + I_w + I_d = 0$$

$$\therefore I_d = -3I_o - (I_g + I_w)$$

$$= -3I_o + I_o\left[\left(\sum_{i=a}^{c} Z_{gi}\right)(Y_{gg} + Y_{wg}) + \left(\sum_{i=a}^{c} Z_{wi}\right)(Y_{gw} + Y_{ww})\right]$$

$$= -I_o\left[3 - \left\{\left(\sum_{i=a}^{c} Z_{gi}\right)(Y_{gg} + Y_{wg}) + \left(\sum_{i=a}^{c} Z_{wi}\right)(Y_{gw} + Y_{ww})\right\}\right]$$

$$\frac{I_g}{I_d} = \frac{\left(\sum_{i=a}^{c} Z_{gi}\right)Y_{gg} + \left(\sum_{i=a}^{c} Z_{wi}\right)Y_{gw}}{3 - \left[\left(\sum_{i=a}^{c} Z_{gi}\right)(Y_{gg} + Y_{wg}) + \left(\sum_{i=a}^{c} Z_{wi}\right)(Y_{gw} + Y_{ww})\right]} = k_1$$

$$\frac{I_w}{I_d} = \frac{\left(\sum_{i=a}^{c} Z_{gi}\right)Y_{wg} + \left(\sum_{i=a}^{c} Z_{wi}\right)Y_{ww}}{3 - \left[\left(\sum_{i=a}^{c} Z_{gi}\right)(Y_{gg} + Y_{wg}) + \left(\sum_{i=a}^{c} Z_{wi}\right)(Y_{gw} + Y_{ww})\right]} = k_2$$

$I_g = k_1 I_d$

$I_w = k_2 I_d$

$I_g + I_w + I_d = 1.0$

$(1 + k_1 + k_2) I_d = 1$

$I_d = \dfrac{1}{1 + k_1 + k_2}$: Fraction of total current flowing in earth.

$I_g = k_1 I_d$: Fraction of total current flowing in ground wire 1.

$I_w = k_2 I_d$: Fraction of total current flowing in ground wire 2.

2-1.2. CONDUCTION

Admittance of Ladder Network,

$$Y = G + jB = \dfrac{1}{\left(\dfrac{Z_{Span}}{2}\right) + \sqrt{R_t \times Z_{Span}}}$$

(2-1.8)

Where Z_{Span} = Self Impedance of one span of ground wire with earth return.

R_t = Average Tower Footing Resistance

$Y' = \sum_{i=1}^{n} Y$ n – Number of parallel lines.

The current discharged to the earth by conduction:

$$I_d \left[\dfrac{\dfrac{1}{Y}}{R_g + \left(\dfrac{1}{Y'}\right)} \right]$$

(2-1.9)

Where R_g is the station grounding resistance

I_d Fraction of fault current discharged through the earth by induction.

Earthing of Electrical System – LV & MV System

Chapter 3

3.0 INTRODUCTION

This chapter covers basic principles involved in earthing of LV (Low Voltage – 415V) and MV (Medium Voltgae – 3.3, 6.6, 11kV) systems. Misconceptions regarding interpretations of IE rules are clarified. Special topics like earthing of tank farms, PV solar plants, clean earth for Control & Instrumentation equipment, bus duct earthing and NGR enclosure earthing are covered in the end.

3.1 EARTHING IN LV SYSTEM

For easy conceptualization single phase network is considered. Three case studies are discussed to bring out the points.

3.1.1 Case 1: Source Grounded, Equipment Ungrounded

Refer Figure 3.1 The source AB is grounded through electrode at E_1. Let electrode resistance to earth be 1Ω. C is the load.

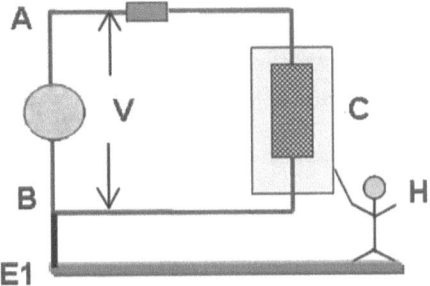

Figure 3.1: Source grounded, Equipment ungrounded

The equivalent circuit is shown in Figure 3.2.

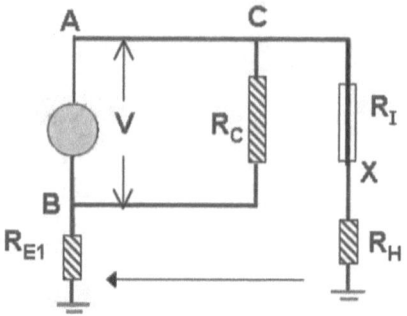

Figure 3.2: Equivalent Circuit

R_C: Equipment load resistance (e.g. $240^2/1000 = 58\Omega$, for 1kW load)
R_H: Human body resistance, say $2k\Omega$
R_I: Equipment insulation resistance
Under healthy conditions, R_I is very high ($M\Omega$). Even if the equipment surface is touched, current through the person is negligible.
Under insulation failure conditions, R_I is zero. Current through the person when he touches the surface of equipment is:

$$I_H = \frac{V}{(R_H + R_{EI})} = \frac{240}{(2000 + 1)} \cong 120 \; mA \tag{3.1}$$

This current is too small for the fuse to operate. But even this small current is high enough to cause injury to a person (Table 2.1).

3.1.2 Case 2: Source Grounded, Equipment Grounded

The source and equipment are grounded through electrodes at E_1 and E_2 (Figure 3.3). Let electrode resistance to earth of E_2 also be 1Ω.

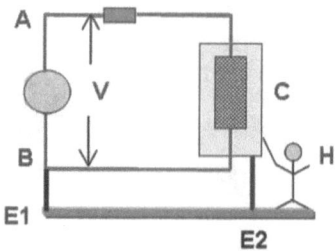

Figure 3.3: Source grounded, Equipment grounded

The equivalent circuit is shown in Figure 3.4

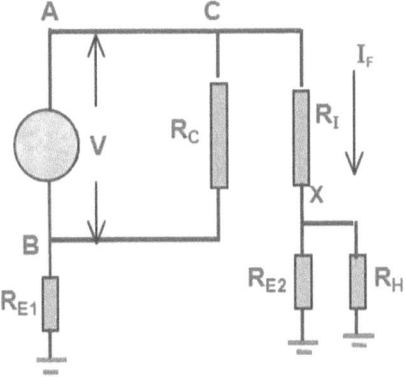

Figure 3.4: Equivalent Circuit

Under insulation failure, $R_I = 0$

$R_{EQ} = R_{E2} // R_H = 1 // 2000 \cong 1\Omega$

$$I_F = \frac{V}{(R_{EQ} + R_{EI})} = \frac{240}{(1+1)} = 120\ A \tag{3.2}$$

$$I_H = \frac{1}{(1+2000)} \times 120 \cong 60\ mA \tag{3.3}$$

The fault current I_F is significant but not high enough and the fuse may or may not trip in desired time. The current through body is less but may still cause injuries.

3.1.3 Case 3: Source Grounded, Equipment Grounded with Bonding Conductor (Figure 3.5)

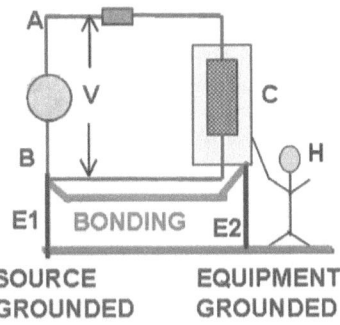

Figure 3.5: Bonding conductor

The equivalent circuit is shown in Figure 3.6.

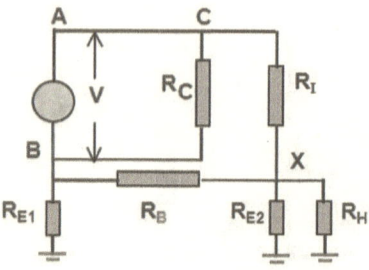

Figure 3.6

It is same as Case 2 but with bonding conductor running between equipment and source. Bonding establishes direct metallic connection between equipment and source. The resistance of bonding conductor R_B is small. In case of insulation failure, majority of fault current returns to source through the metal and negligible current through earth ensuring human safety. Also the fault current magnitude is significantly higher due to low resistance metallic return path and positive fault isolation by fuse or circuit breaker is ensured.

The increased safety offered by widely used 'Three Pin Socket' is obvious now. One of the pins is used for direct metallic connection between source and load. Refer Figure 3.7. In the top figure Earth Continuity Conductor is absent. In case of insulation failure, the current returns through body. This is unsafe earthing. In the bottom figure Earth Continuity Conductor is present ('third wire'). In case of insulation failure, the current returns through the metallic third wire. This is safe earthing.

Unsafe Earthing
No Earth Continuity Conductor

Safe Earthing
With Earth Continuity Conductor

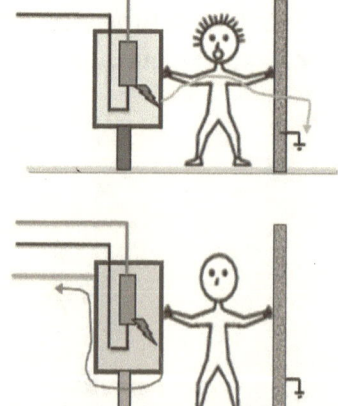

Figure 3.7

The above case studies illustrate an important concept that in LV (and also MV) systems, for getting the fault current back to the source, 'mother earth' should not be dependent upon. The fault current shall be carried back to the source through metallic connection (e.g. cable armour, dedicated earth strips say 25 x 3 mm, etc). Since no current is injected into the earth, touch and step potentials are irrelevant in these cases.

In fact, as per NEC, Sec 250.4 (A) (5), "the earth shall not be considered as an effective ground-fault current path". Earth can be only supplementary to a metallic return path for flow of current back to source neutral.

Even the Indian Standard (IS 3043) recognizes this fact. In Cl 0.3, It mentions that 'As a matter of fact, the earth now rarely serves as a part of the return circuit, but is being used mainly for fixing the voltage of the system neutrals'

3.1.4 Earthing in TPN System

This discussion is applicable only in industrial plant environment and not in utility distribution system. Two alternatives are shown in Figure 3.8 and 3.9. Single phase loads may be present at MCC and PCC but are not shown explicitly in figures.

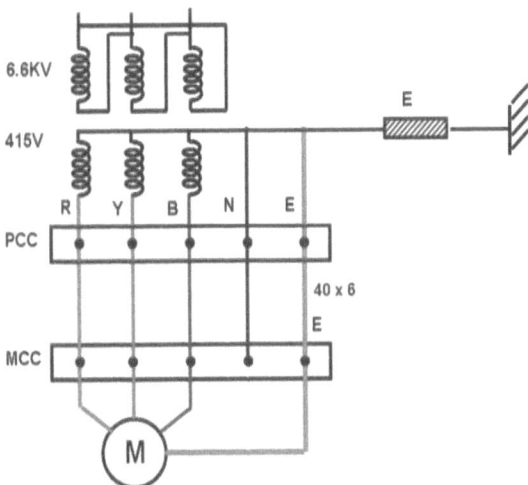

Figure 3.8: Alternative 1

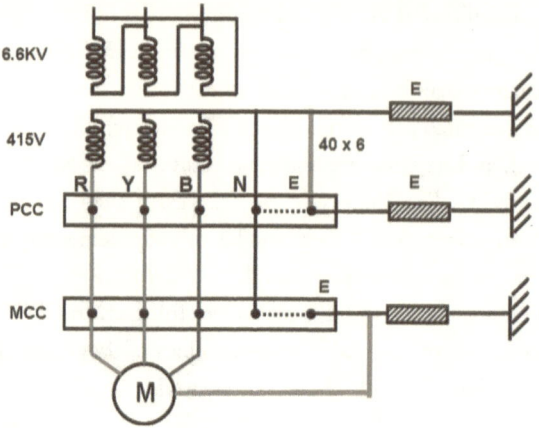

Figure 3.9: Alternative 2

In Alternative 1, the basic premise is that neutral must carry only (steady state) unbalance current whilst the earthing conductor shall carry only ground fault current. Neutral and earthing conductor shall not be interconnected at any place except at service entrance (neutral of feeding transformer). Consider three LT switchgears A, B and C with A feeding B and B feeding C. Metallic earthing conductor must run between C to B, B to A and A to service neutral. For this reason, in literature it is sometimes referred as 'earth continuity conductor'.

In Alternative 2 Neutral and Protective earth wire are connectd to each other and bonded to earth.

Alternative 1 as depicted in Figure 3.8 is rarely achieved in field. The earthing conductors (e.g. 40x6 mm GI strips) can run along cable trays from MCC to PCC to Transformer and the cable tray support structures themselves are earthed. Only difference is that unlike Alternative 2, no intentional earthing arrangement is made at local PCC or MCC.

For positive clearing of earth faults, it must be ensured that armour of all cables emanating from MCC and PCC are firmly bonded to earth bus of PCC and MCC. It is to ensure that ground fault current returns to source (transformer) neutral via metal.

3.1.5 Earthing in Utility Distribution System in Urban Areas

A typical LT distribution schematic is shown in Figure 3.10. LT power supply is derived from Distribution Transformer (DT) fed by 11kV RMU (Ring Main Unit). LT Panel (LTP) is switchgear typically with 1 incomer with 8 to 10 outgoing circuits to Feeder Pillars (FP). The Feeder Pillars in turn supplies power to Mini Pillar (MP). The Meter Cabin (MC) in ultimate customer premises is fed by Mini Pillar. Typical cable sizes used in different segments are indicated in Figure 3.10. The cables are 4 cores (3 phase and Neutral) with armour. Armour is also referred as PE (Protective Earth).

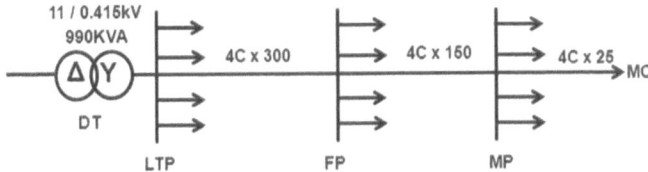

Figure 3.10

Depending on type of bonding between Neutral and armour, different classifications are defined in IEC 60364-1 [30].

3.1.5.1 TN-S: Separate Neutral Conductor & Protective Earth(PE)conductor through out the System

Refer Figure 3.11. Neutral and PE are bonded only at source. This is rarely adopted in urban distribution system spread over wide area. Any damage to PE due to external intrusion can result in open circuit and this will result in break of metallic path for return of fault current back to source.

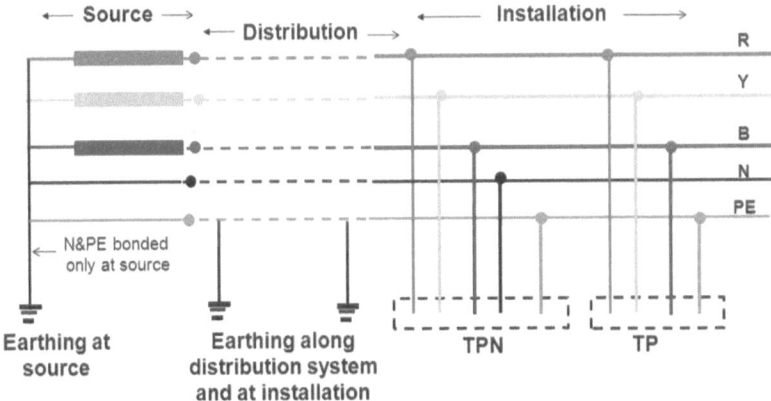

Figure 3.11: TN-S

3.1.5.2 PME system – Protective Multiple Earthing System

Refer Figure 3.12. This is also called TN-C-S - Combined Neutral Conductor and Protective Earth (PE) from source and separated at installation. In urban distribution use of single conductor for both Neutral and PE (4C without armour or 3C with armour) is rarely used.

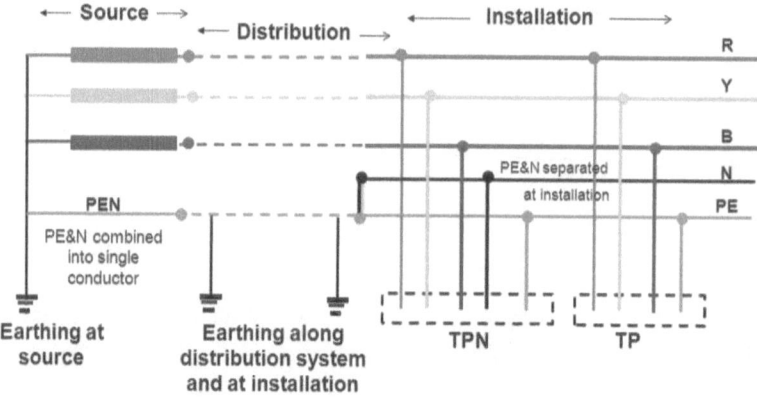

Figure 3.12: TN-C-S (PME)

3.1.5.3 Modified TN-C-S

Refer Figure 3.13. Neutral Conductor & Protective Earth (PE) are bonded at source, along route and at installation. This is the preferred method in urban distribution. Cable used is 4C with armour. Even if armour or neutral breaks creating open circuit at that location, alternate path is available for fault current to return to source via metal as PE (armour) and neutral are bonded at intermediate locations. For example, in Figure 3.10, Neutral and Armour are bonded and earthed at LTP, FP, MP and MC.

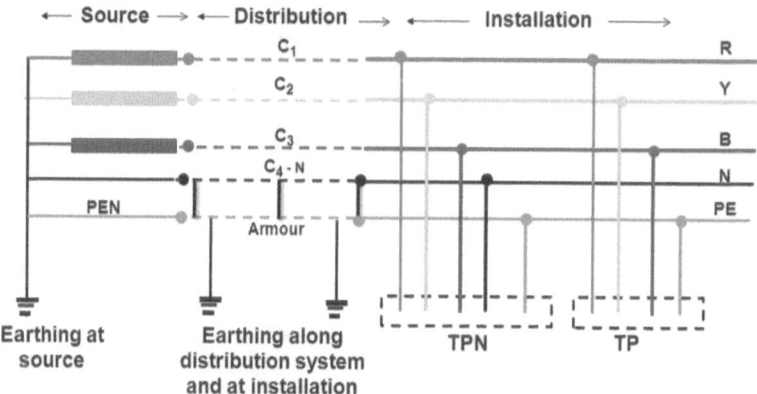

Figure 3.13: Modified TN-C-S

3.1.5.4 TT System

Refer Figure 3.14. There is no metallic connection between source earthing and installation earthing. This is not preferred. In case of ground fault in installation, fault current has to return via earth only and may result in touch potential that is not safe.

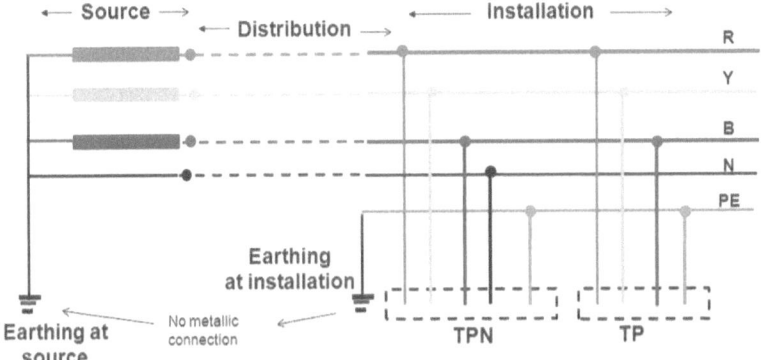

Figure 3.14: TT System

3.1.5.5 Touch Voltage Due to Broken Conductor

Refer Figure 3.15. Consider a street lamp rated for 2 x 400W. Assume the resistance of local earth electrode is 25Ω (moderate value).

1φ Load of 2 x 400 = 800W

Phase Voltage = 240V

Load Resistance

$$R_L = \frac{240^2}{800} = 72\Omega \tag{3.4}$$

In case of broken conductor in return path, the resulting current

$$I = \frac{240}{97} = 2.47A \tag{3.5}$$

Voltage across Load V_L = 2.47 x 72 = 178V

Touch Voltage V_T = 2.47 x 25 = 62V

Safe touch voltage as per Indian Standards is 65V (in some European countries, even upto 150V is permissible). If 65V is taken as limit, the present case is only marginally safe.

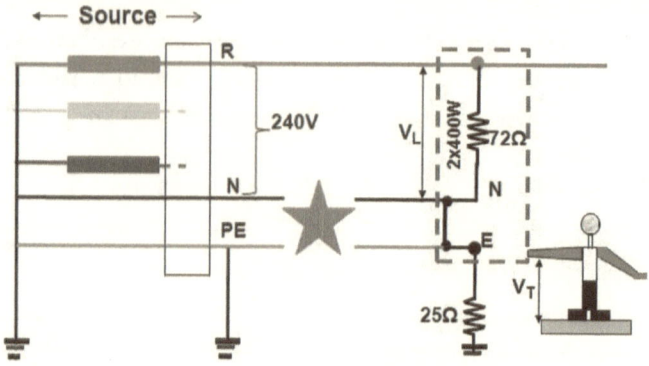

Figure 3.15: Earth Electrode Resistance – Moderate

Let us repeat the above case with a local earth electrode of higher resistance of 100Ω. Refer Figure 3.16.

In case of broken conductor in return path, the resulting current

$$I = \frac{240}{172} = 1.4A \tag{3.6}$$

Voltage across Load V_L = 1.4 x 72 = 100V

Touch Voltage V_T = 1.4 x 100 = 140V

The touch voltage is much higher than allowable voltage of 65V and is very unsafe. This brings out once again the importance of metallic path for the current to return to source neutral instead of depending on 'mother earth'.

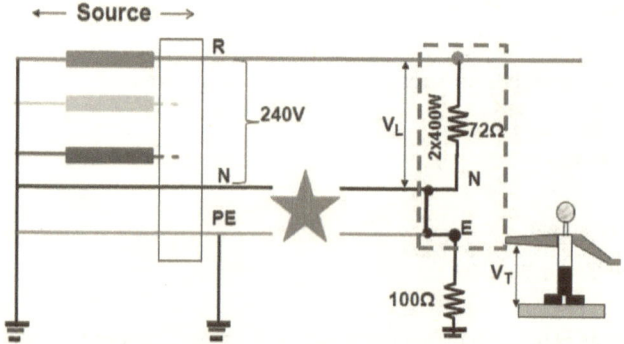

Figure 3.16: Earth Electrode Resistance – High

It is interesting at this juncture to point out advantage of using LEDs for street lighting. The input impedance in case of LED is high (current drawn is less than 1A) and hence variation of electrode resistance has only marginal impact on touch voltage as well as lumen output in case of broken conductor.

3.1.5.6 Single Core Cable Armour Bonded at Only One End

Only one end of armour (usually sending end) is earthed and the other end is insulated. This is called *single point bonding*. Refer Figure 3.17.

If both ends are bonded (*solid bonding*) there will be circulating current in armour resulting in increased heat generation. In this case cable has to be derated.

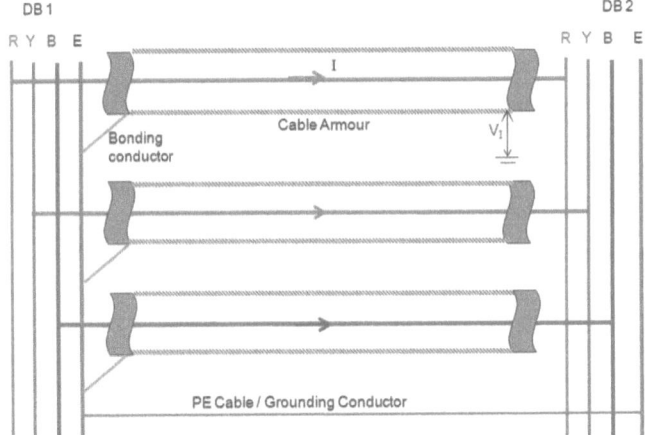

Figure 3.17: Single point bonding

Single point bonding prevents circulating current flow in armour. However, in this method, the free end of the armour (insulated) would develop induced voltage V_I. Indian Electricity Rules permit 65 volts as the limit of such induced voltage. Voltage induced in armour is determined by armour diameter, spacing between cables (trefoil or flat formation) and phase currents. For LV and MV cables, induced voltage in armour is approximately given by $V_I \cong 55mV/Amp/KM$.

For example, for a current of 750A and cable length of 0.5KM, induced voltage in armour = 0.055 x 750 x 0.5 = 21V.

Unlike solid bonding, single point bonding creates discontinuity in armour circuit and inhibits flow of fault current returning back to source via a metal. In these cases, it is mandatory to provide additional grounding conductor between two distribution boards connected by single core cables. Refer Section 5.4.3 of IEEE Std 575 [25].

In case of earth fault within single core cable (X), the fault current can return via the armour (Figure 3.18). In case of an earth fault in any outgoing feeder of the receiving end distribution board (Y), the separate ground conductor facilitates return of the earth fault current through the metal to the upstream source [14].

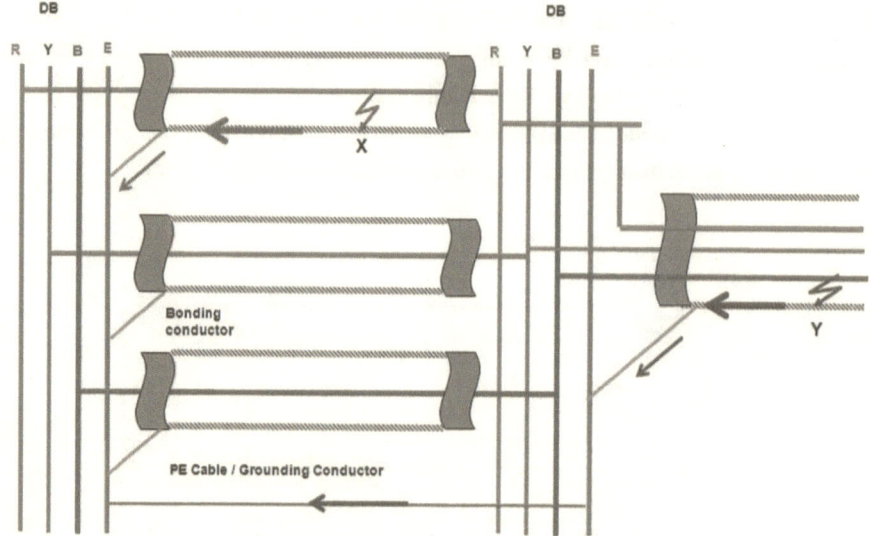

Figure 3.18: of Fault current distribution

3.2 EARTHING IN MV RESISTANCE GROUNDED SYSTEM

Here also the earthing conductor must run all over the concerned area. Armour of cables and equipment enclosures are firmly bonded to earthing conductor which can be typically 65 x 10 mm GI strip. This is to ensure that fault current returns to Neutral Grounding Resistor (NGR) only through a metallic path. Refer Figure 3.19.

In power plants and industrial plants ground fault current is limited to typically 300A to 400A. In mining industry, due to safety considerations ground fault current is limited to less than 50A. Ground fault detection is a major issue in the later case. For positive relay pick up during a ground fault for selective isolation, it is essential that current returns through metal rather than earth.

More details on this topic are covered in Sec 5.9.2.

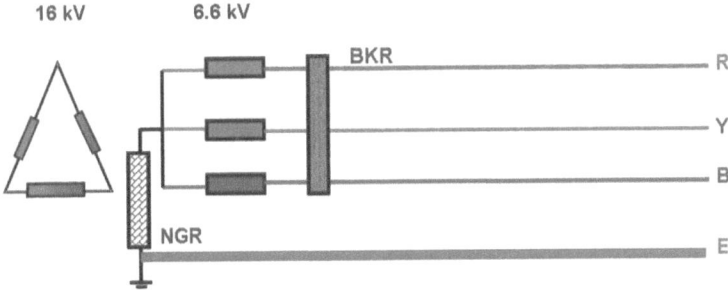

Figure 3.19: Earthing in MV System

3.3 EARTHING IN MV UNGROUNDED SYSTEM

At first it looks strange to talk of earthing conductor in ungrounded system. Two motor feeders fed by ungrounded system are shown in Figure 3.20.

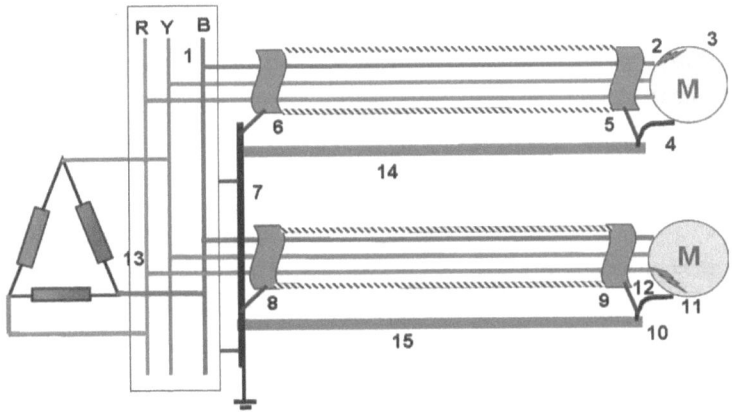

Figure 3.20: Earthing in Ungrounded System

Of course for one ground fault, no current flows. Assume two ground faults, on Phase R in one motor and on Phase B in another motor. Under ideal conditions it must lead to flow of large phase to phase fault current to be sensed by over current relays leading to tripping of both motors {refer (Cl 4.6(v)}. Without earthing conductor, the path taken will be (through cable armour) from 1 to 13 creating a phase to phase short circuit. If the armour can't sustain the flow of fault current, it will be damaged and can result in open circuit. If the cable does not have armour (unarmoured cable), sufficient phase to phase fault current may not flow as the return path is only through earth. On the other hand if earth continuity conductor is provided, the path taken will be [1 to 4, 14, 7, 15, 10 to 13]. This facilitates flow of fault current between phases without impediment that improves relaying performance.

3.4 MISCELLANEOUS TOPICS

3.4.1 (Mis) Interpretation of IE Rules for Transformer Neutral and Body Earthing

IE rules 1956 – 2000 [22] is used for certifying the safety of electrical installations. One of the rules 'quoted' often is:

> "Neutral of transformer shall be connected to *two separate earth electrodes*. Similarly body of transformer shall be connected to *two separate earth electrodes*" - This is *not* a correct interpretation of IE rules.

The rule states as follows:
Cl 67/61 Connection with earth

> "The neutral point of every generator and transformer shall be earthed by connecting it to the earthing system as defined in Rule 61(4) and hereinabove by not less than *two separate and distinct connections*".

> "The frame of every generator, stationary motor, portable motor, and the metallic parts (not intended as conductors) of all transformers and any other apparatus used for regulating or controlling energy and all medium voltage energy consuming apparatus shall be earthed by the owner by *two separate and distinct connections with earth*".

The emphasis of the above rule is to ensure 'last mile connectivity' by metal. Two separate and distinct 'connections with earth' ensures that even if one connection is broken, metallic connection to earth grid is still available. Interpreting that two separate and independent earth electrodes are required as per IE rules is fallacious. Two isolated earth electrodes can't match the performance of entire earthing grid. Resistance of isolated electrode is of the order of 20 to 30Ω whilst that of earthing grid of switchyard or plant is less than 0.5Ω. In summary, what is required by IE rules is connection by two independent strips or cables from neutral or body to earth grid. If neutral or body is connected to two separate earth pits, it is essential that these earth pits in turn are connected to earthing grid.

3.4.2 (Mis) Interpretation of IE Rules for LA Earthing
Cl 92(2) of IE rules states the following:

> "The earthing lead for any lightning arrestor shall not pass through any iron or steel pipe but shall be taken as directly as possible from lightning arrestor to a separate earth electrode and/or a junction of earth mat already provided for high and extra high voltage sub-station subject to the avoidance of bends wherever practicable."

Two separate earth electrodes are not mandatory. Two separate connections from LA to earthing grid are *also acceptable*. Only caution is that the lead length must be as short as possible without sharp bends. For this reason usually two treated earth pits are created very near to LA. The function of earth pits is to drain the electric charges from lightning strike to ground effectively. Step and touch potentials are irrelevant in this case as the current duration is of the order of microseconds. The earth pits created for LAs shall also be connected to main earthing grid which will reduce overall earth grid resistance.

The down leads from LA to earth pass through Surge Counters. Care must be taken during mounting of Surge Counters on structures. Surge Counters must be insulated from support structures. Otherwise surge may bypass Surge Counter.

3.4.3 Tank Earthing

Metallic tanks are installed to store fuels and chemicals in power plants and refineries. The tank body should be connected to station earth grid by multiple connections. Minimum four connections are provided at corners for smaller tanks. For large tanks, spacing between two earth connections shall not exceed 30 meters along tank perimeter. If connections are extended to separate earth electrodes, then the electrodes should be interconnected at below ground level. This interconnected earth electrode system for tanks should also be connected to the plant earth grid. Metal objects like pipes and rails within the vicinity of tank farm shall also be bonded to earth grid.

It is not recommended to provide any other lightning protection and separate earth pits for the same. No lightning spikes should be provided on fuel tanks. They tend to attract the lightning towards the tank and there is probability of shielding failure that may result in direct hit to the tank. No spike is lesser evil. A properly earthed tank, as described earlier, is considered to be self protected against lightning.

3.4.4 Bus Duct Earthing

In power plants and industrial plants, bus ducts are used as incomers from transformer to switchgear when current ratings are high (say above 3000A). In case of LV system (415V), the transformer neutral is solidly grounded. In case of MV system (3.3, 6.6, 11kV), the transformer neutral is usually grounded through NGR (Neutral Grounding Resistor). When any earth fault occurs within the bus duct, it is essential that the current returns back to source via the metal for positive operation of protection system. Bus duct enclosures may have discontinuity and should not be depended for earthing connection. To achieve this, each busduct shall have an earth bus running the full length on the outside. The switchgear shall have a provision to extend the switchgear earth bus to the Busduct terminal flange. At the transformer end, the earth bus of the Bus duct shall be earthed, with a direct earth flat to the earth grid (thereby joining the transformer neutral (for LV) and transformer NGR (for MV). For LV busducts 65x10 mm GI and

HV busducts 50x6 mm GI strips will suffice. Typical arrangement for LV Busduct (TPN) is shown in Figure 3.21.

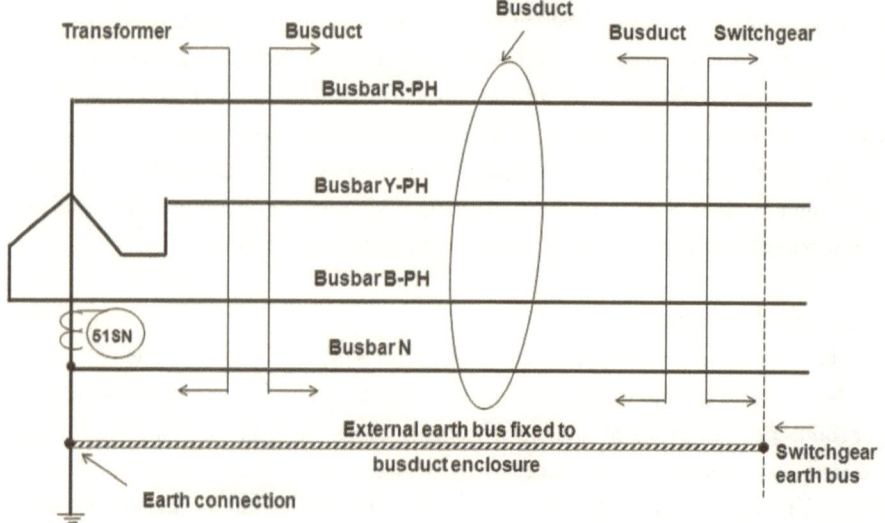

Figure 3.21

3.4.5 Clean Earth

Earthing of equipment in DCS or SCADA is a specialized subject. The proper grounding of sensitive electronic equipment is critical in achieving the reliable performance of these systems. The term 'electronic' is used for micro processor based systems such as computers, PLC, industrial/power plant DCS, telecommunication systems, medical diagnostic, imaging systems etc.

The functions of major connections/wire in a typical DCS panel (Figure 3.22) are as follows:

(i) Phase Wire (P): This is from AC source usually from UPS.

(ii) Neutral Wire (N): Provides return path for the current back to source neutral. Neutral size shall be minimum same size as Phase.

(iii) Power Earth Bus (PG): This is also termed as 'Power Ground Bus'. All non current carrying parts are bonded to Power Ground Bus. It directly rests on panel. In general earth strips, say 50x6 mm GI strips, run all over the plant including room housing DCS panels. The Power Earth bus in individual panel is connected to the earth strip at the nearest location. The panel body is also connected to earth strip. The earth strip is connected to 'Power Earth Pit' to which source neutral is also connected. Neutral and power earth are bonded only at the source. In some literature, 'power earth' is also termed as 'dirty earth'.

(iv) Signal Earth Bus (IG1): It can be termed as 'Instrumentation Ground 1'. The electronic equipment may have different DC voltage systems like +12/0/-12, +5/0/-5. Usually 0 of DC control supply is connected to 'Signal Earth Bus' to obtain stable reference voltage in all DCS panels. Signal Earth Bus (IG1) is insulated from panel body. Signal Earth Bus is looped from panel to panel using insulated copper wires and finally terminated on 'Clean earth Pit'. 'Clean Earth' is also sometimes referred as 'Electronic Earth'.

(v) Sheath Earth Bus (IG2): It can be termed as 'Instrumentation Ground 2'. Control cables from field instruments to DCS panels are multi-core twisted pair copper cables with shielding wires (Aluminum mylar) termed as 'sheath'. The sheath of individual cable terminating in DCS is connected to Sheath Earth Bus. IG2 is also insulated from panel body. The sheath earthing is done only at one end, generally at panel end. Sheath Earth Bus is looped from panel to panel using insulated copper wires and finally terminated on 'Clean earth Pit' as mentioned in section (iv) above or in some cases on a separate 'clean earth pit'. The shield earth is provided to protect the I&C signals carried by cables from field instruments to control room from induced voltages and external interferences. Sheath (screen) has to be present from termination to termination. If portion of sheath is removed during field wiring, there is a possibility of EMI from neighbouring cables. If muti core cable is used, each twisted pair shall be individually screened and overall screening covering all the pairs shall be provided.

Minor variations in arrangement from that shown in Figure 3.22 are possible depending on recommended practices adopted by individual vendor and or user. For example, some prefer to keep 'Clean Earth Pits' isolated from "power earth or dirty earth'. The connection shown in dotted rectangle in Figure 3.22 is absent. The resistance of 'clean earth pit' is designed for less than 0.5Ω. This may be difficult to achieve and maintain in the long run in practice. To ensure low earth pit resistance, these earth pits need to be periodically treated preferably during unit outages. The majority opinion is to interconnect the 'clean earth' with 'power earth' as shown in Figure 3.22. The 'power earth pit' in turn is connected to 'main earthing grid or station earth mat' whose resistance is very small (less than 0.2Ω).

In some projects IG1 and IG2 are merged into a single bus.

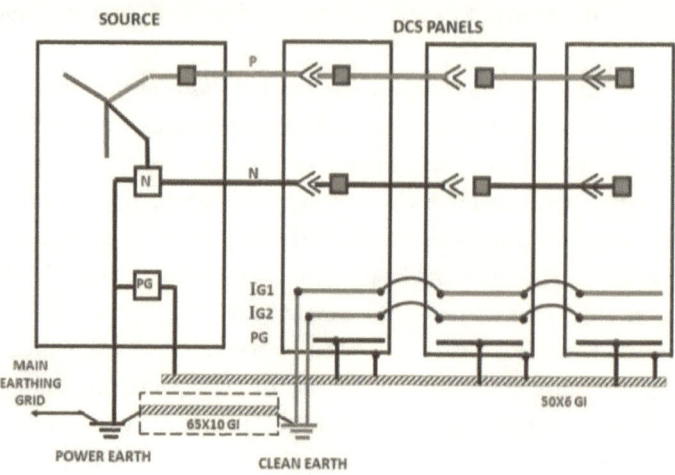

Figure 3.22

3.4.6 Solar Farms

In utility scale PV plants, solar field is spread across a huge area. The Solar PV modules are mounted on structural steel members, which are embedded in the ground and technically behave as earth electrodes. These steel structures along with Module frames are connected with each other and bonded to the buried strip. For utility scale PV plants 25 x 3 mm GI strip buried at a depth of 600 mm is adequate. PV Cell is current limited device (Figure 3.23). The maximum fault contribution is limited by the galvanically coupled cluster of modules and is generally less than 2kA. A 25x3 mm GI strip can carry upto 6kA for 1 sec (80A/mm^2 as per Table 1.3). As the area enclosed as well as length of buried conductor is very large in solar farm, earth grid resistance is well below desired value of 1Ω.

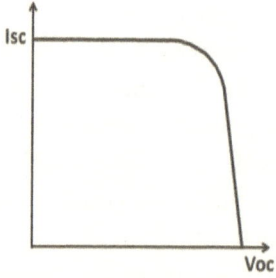

Figure 3.23

For lightning protection, Early Steamer Emission Air Terminals (ESEAT) are widely used. With a few ESEATs solar farms of large area can be protected. For example an ESEAT 600 mm long mounted on a 5meter mast on building housing inverters can offer protection radius of 100 meters. Each ESEAT air terminal is connected to redundant earth pits using two numbers single core (50 mm²) cable of copper or aluminum. Resistance of individual earth pit per se is not influencing factor as it has to only drain the lightning charges to ground. The earth pits in turn are also connected to earth grid of solar farm to provide more paths for draining the charges.

In poly crystalline modules it is recommended to earth the -ve of DC system at the inverter end (Figure 3.24). This is not a protective earthing. This is done to mitigate the phenomena of PID (Potential Induced Degradation) of PV modules.

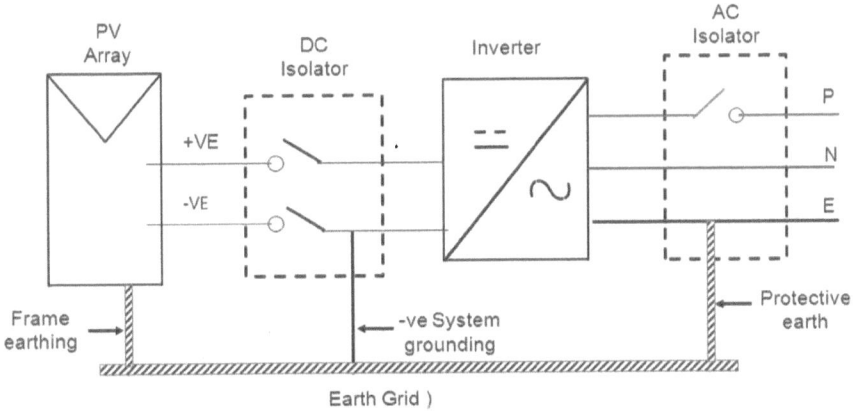

Figure 3.24

3.4.7 NGR Enclosure Earthing

The sheet steel enclosure housing Neutral grounding Resistor stack is on insulated base as shown in Figure 3.25.

In conventional enclosures two earthing terminals are usually provided for connection to nearby earthing strip connected to main plant earthing grid. Unlike conventional practice, the enclosure of NGR *should not be earthed*. It is recommended to manufacture enclosure for NGR without earthing terminals so that enclosure can't be earthed even by mistake.

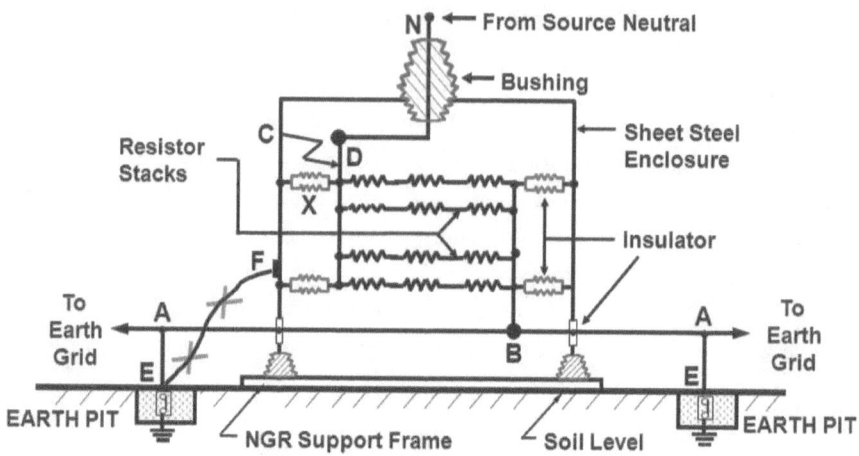

Figure 3.25: NGR Cubicle Earthing

3.4.7.1 Analysis

NGR is mounted on insulators. The resistance stacks rest on support insulators mounted inside enclosure. The danger posed by earthing the enclosure in case of failure of one of the support insulators is explained below.

Case 1

Connection EF is absent (enclosure not earthed). All insulators are healthy. For a ground fault, the return path is E-A-B-Resistor-D-N to source neutral.

Case 2

Connection EF is absent (enclosure not earthed). Assume that insulator X has cracked. For a ground fault, the return path is still E-A-B-Resistor-D-N to source neutral.

Case 3

Connection EF is present (enclosure earthed). Assume that insulator X has cracked. CD is conducting path. For a ground fault, the return path will be E-F-C-D-N to source neutral. The resistor is completely bypassed. Instead of resistance grounded system, it has become solidly grounded system with disastrous consequences from core damage point of view. Significance of core damage in rotating machines is covered in Sec 5.8.2.

Under normal operating conditions NGR carries very little current. IR testing of NGR is rarely carried out as part of routine maintenance practices. Hence a leaking insulator may be present but its effect is felt only during fault conditions.

The practicing engineer's anxiety regarding safety when the enclosure is not earthed can be addressed as follows. When an earth fault occurs, if the enclosure is not earthed, the enclosure may experience a rise in voltage. But modern protection systems clear the faults within one second. Hence the probability that someone touches the enclosure exactly during that one second when the fault has occurred is very remote. As extra precaution, the enclosure is kept within a fence or located at elevated platform. Refer Sec 5.9.2 for more details on this topic.

Grounding of Electrical System – Ungrounded System

Chapter 4

4.0 INTRODUCTION

The two areas which have significant impact on system protection and operation are earthing and grounding. Earthing of electrical system has been dealt in detail in Chapters 1 to 3. In Chapters 4 to 7, concepts of grounding will be developed.

4.1 UNGROUNDED VS GROUNDED SYSTEM

Up to 1940, most of the systems were operated as ungrounded system. The neutrals of the system were kept floating as the ground connection is not useful for transfer of three phase power in three-wire system. The majority of the faults (70%) on any system are line to ground faults. In ungrounded system, due to absence of return path, the ground fault current is very low. As the service is not interrupted, the fault can be located and rectified at leisure. But soon problems like transient over voltages and arcing grounds began to surface resulting in insulation failure. To overcome these problems, grounding the neutral was considered as a possible solution. Of course the grounded system results in flow of large ground fault current. The over voltage problem in ungrounded system is replaced by over current problem in grounded system. Even today, this debate on grounded vs. ungrounded system is going on and 'The Preferred Alternative' does not exist!

4.2 DIFFERENCE BETWEEN NEUTRAL AND GROUND

Before we proceed further, the concept of neutral will be introduced [5]. The neutral and ground are not always the same. The ground is always at zero voltage plane, whereas the neutral can be at ground (zero) potential or it can have some non - zero potential with respect to the ground. In case the neutral is at zero potential, then there is no neutral shift. If neutral is at non - zero potential, then it is considered to be shifted.

For definition of neutral refer Figure 4.1. Connect three ideal and equal resistors to three phases at point of interest. The common junction point is the 'neutral'. The voltage of common junction point with respect to ground is neutral voltage. From this definition follows an important fact – 'The zero sequence voltage at any point in network corresponds to the neutral shift at that point in the network'.

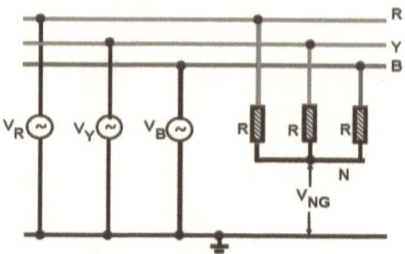

Figure 4.1: Neutral and Zero Sequence Voltage

From Figure 4.1

$$\left[\frac{(V_R - V_{NG})}{R}\right] + \left[\frac{(V_Y - V_{NG})}{R}\right] + \left[\frac{(V_B - V_{NG})}{R}\right] = 0$$

$$V_R + V_Y + V_B = 3V_{NG} \tag{4.1A}$$

From the theory of symmetrical components,
$$V_R + V_Y + V_B = 3V_0 \tag{4.1B}$$
Hence, $V_{NG} = V_0$ (4.1C)

Open delta PT connection described in Sec 4.9 uses the above concept to measure neutral shift.

4.3 UNGROUNDED SYSTEM (BALANCED OPERATION)

In ungrounded system, there is no intentional connection to ground provided for exclusive grounding purpose. In reality, however, no ideal ungrounded system exists, since phases get coupled to ground through stray capacitances (Figure 4.2).

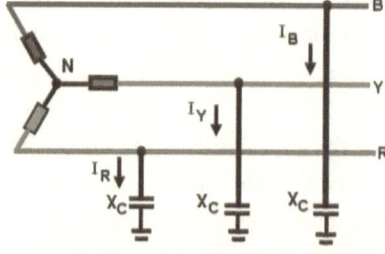

Figure 4.2: Ungrounded system – Balanced Steady state operation

The phasor diagram is shown in Figure 4.3.

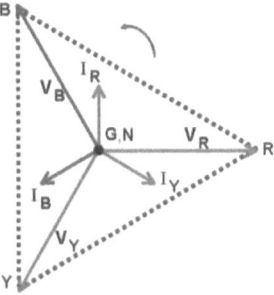

Figure 4.3: Phasor diagram

The phase voltages are V_R, V_Y and V_B. (It is common to write with single subscript for voltage to ground, e.g. V_R instead of V_{RG}). The line voltages are V_{RY}, V_{YB} and V_{BR}.
$|V_R| = |V_Y| = |V_B| = 1\text{pu}$ (say $11\text{kV}/\sqrt{3}$)
$|V_{RY}| = |V_{YB}| = |V_{BR}| = \sqrt{3}\text{pu}$ (i.e. 11kV)

The phase current leads phase voltage by 90°.
$I_R + I_Y + I_B = 0$
$V_R + V_Y + V_B = 0$

From Eqn (4.1),
$V_{NG} = V_0 = 0$

Neutral is at ground potential and there is no neutral shift.

4.4 UNGROUNDED SYSTEM (FAULT CONDITION)

Now consider a line to ground fault on phase R (Figure 4.4). The phasor diagram is shown in Figure 4.5. R phase capacitance is shorted since fault is on phase R. Now R is at ground potential G. Voltages that appear across other two capacitances are line voltages V_{YR} and V_{BR}.

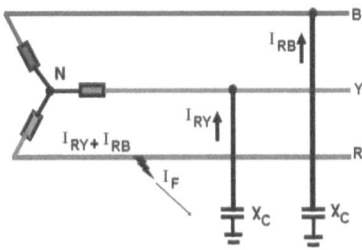

Figure 4.4: Ungrounded system – Line to ground fault

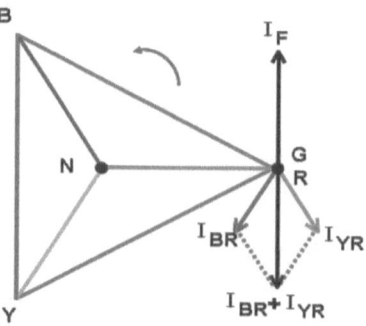

Figure 4.5: Phasor Diagram

The capacitive currents lead respective voltages by 90°. There is no change in line voltages but the voltages to ground of unfaulted phases rise to line voltage.

$|V_{RY}| = |V_{YB}| = |V_{BR}| = \sqrt{3}pu$

Line voltage triangle is still equilateral.

$|V_Y| = |V_B| = \sqrt{3}pu$

The fault current can be calculated as follows:

$|I_{BR}| = |I_{YR}| = \sqrt{3}/X_C$

$I_F = I_{RB} + I_{RY} = 3/X_C$ (4.2A)

The capacitive current per phase ($1/X_C$) of typical industrial system or auxiliary system of a power plant is from 2 to 5A. Thus the fault current will be in the range of 10A. Since the current is capacitive, voltage is maximum at current zero. If the fault is transient, because of repeated strikes due to capacitive nature of current, voltage build up might occur leading to insulation failure. This is called 'arcing faults'. However in practice these are rare occurrences and hence will not be discussed further.

The open delta voltage (Figure 4.6) is given by

$V\Delta = V_R + V_Y + V_B = 3V_{NG}$ (4.2B)
$V_R = 0$ (4.2C)
$|V_Y| = |V_B| = \sqrt{3}$ (4.2D)
$V\Delta = 3pu$ (4.2E)

From Eqn. (4.1C), (4.2B) and (4.2E)

$V_{NG} = V_0 = 1pu.$ (4.2F)

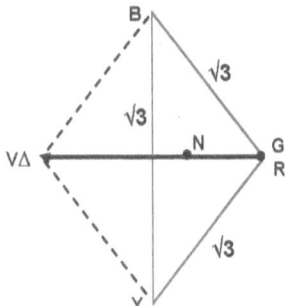

Figure 4.6: Open delta voltage

The neutral is no more at ground potential but is shifted by as much as phase voltage. For example, in a 11kV system, at the point of fault, the neutral voltage instead of zero will be at $11/\sqrt{3}$kV. The corresponding open delta voltage will be $11 \times \sqrt{3}$kV.

It will be educative to work out sequence voltages for the above condition. Using phasor notation, from Figure 4.5 and Eqns. (4.2B to 4.2D)

$V_R = 0$
$V_Y = \sqrt{3}\ L{-}150°$
$V_B = \sqrt{3}\ L{+}150°$

From theory of symmetrical components,

$V_0 = \dfrac{1}{3}[0 + \sqrt{3}L-150 + \sqrt{3}L150]$

$\quad = -1$

$V_1 = \dfrac{1}{3}[0 + aV_Y + a^2V_B]$

$\quad = \dfrac{1}{3}[1L120 \times \sqrt{3}L-150 + 1L-120 \times \sqrt{3}L150]$

$\quad = \dfrac{1}{3}[\sqrt{3}L-30 + \sqrt{3}L30]$

$\quad = 1$

$V_2 = \dfrac{1}{3}[0 + a^2V_Y + aV_B]$

$\quad = \dfrac{1}{3}[1L-120 \times \sqrt{3}L-150 + 1L120 \times \sqrt{3}L150]$

$\quad = \dfrac{1}{3}[\sqrt{3}L-270 + \sqrt{3}L270]$

$\quad = 0$

Since V2 is zero, Negative sequence voltage can't be used as a handle to defect faults in ungrounded system. Zero sequence voltage, derived from open delta PT connection (Figure 4.14), is generally used in these cases.

For comparison, we will evaluate the neutral shift in solidly grounded system. The phasor diagram, for line to ground fault on phase R, is shown in Figure 4.7. There is no change in line voltage V_{YB} as fault is on phase R.

$|V_{YB}| = \sqrt{3}$pu; $|V_{RY}| = |V_{BR}| = 1$pu

Line voltage triangle is isosceles.

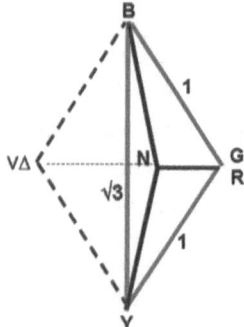

Figure 4.7: Phasor diagram – Solidly grounded system

The voltages to ground of unfaulted phases remain at 1pu compared to $\sqrt{3}$pu in ungrounded case.

$V_R = 0$; $|V_Y| = |V_B| = 1$;
$V\Delta = V_R + V_Y + V_B$
$= 3V_{NG} = 1$pu
$V_{NG} = V_0 = 0.33$pu

This brings out an important fact that neutral shift *occurs irrespective of type of grounding*. In case of ungrounded system neutral shift is high (1pu) and in solidly grounded system neutral shift is low (0.33pu). A sensitive voltage relay connected across open delta PT can be used to detect ground fault even for solidly grounded system.

4.5 ADVANTAGES OF UNGROUNDED SYSTEM

Service continuity, even with one ground fault hanging, is feasible. This is desirable in continuous process industries.

The fault current at the point of fault is very low. The core damage in rotating machine (discussed in Sec 5.8.2) is minimal.

4.6 DISADVANTAGES OF UNGROUNDED SYSTEM

There is a possibility of restrikes when opening capacitive currents of 10A to 15A. This leads to voltage build up and insulation failure.

During ground faults there is over-voltage on healthy phases and neutral shift is high. This results in following design considerations:

(i) Neutral has to be fully insulated.

(ii) Lightning arrestor has to be rated for 100%. For example, in a 11kV system, LAs have to be rated for 11kV even though they are connected from phase to ground.

(iii) The over voltage factor for PT shall be 1.9 pu (instead of 1.5 pu)

(iv) Even in ungrounded system SOP (Standard Operating Procedure) has evolved to detect, locate and isolate the earth fault within a reasonable time, typically 8 hours. This is to prevent multiple earth faults hanging in the system. The cables are rated for full line voltage. For a 11kV system, the rating shall be 11kV/11kV, i.e. phase to phase insulation will be 11kV and phase to ground insulation will also be 11kV. This is called UE grade (Unearthed grade) as per IS 7098-2 and Category C as per Cl 4.1 of IEC 60502-2. Instead of 11kV/11kV UE grade, next higher size Earthed grade cable (12.7kV/22kV) can also be used. Refer Sec 5.13 for more discussions on related topic.

(v) The chances of multiple ground faults hanging on the system are high unless prompt action is taken to identify and isolate the first ground fault. In this context, it is useful to recognize following trip logic. Refer Figure 4.8.

(a) Fault on phase R to G (Ground) and phase Y to G on Feeder A: Feeder A trips.

(b) Fault on R to G on feeder A and fault on Y to G on feeder B: Both feeders A and B trip.

(c) Fault on R to G on feeder A and fault on R to G on feeder B: None trips; case of multiple ground faults hanging.

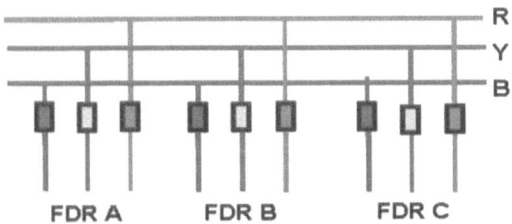

Figure 4.8: Ground fault detection

4.7 GROUND FAULT DETECTION IN UNGROUNDED SYSTEM

In ungrounded system current handle for ground fault detection is not reliable as fault current is too low. Only voltage signal is available for ground fault detection. It may be emphasized that voltage can be used for ground fault *detection* but not for *location*. The difference between voltage handle and current handle is that voltage is a system wide attribute while current is feeder specific. In Figure 4.8, voltage is same for feeder A, B or C. In fact it is the common 'bus' voltage. The currents are however feeder specific. In case of grounded system, if a ground fault occurs on feeder A, large current flows on feeder A and this current handle is used to trip feeder A. Location of ground fault that fault is only on feeder A is possible due to availability of sufficient current handle. In case of ungrounded system this is not available. Voltage handle indicates the presence of ground fault. To identify the location, feeders are tripped one by one. When the faulted feeder is tripped voltage handle indicates healthiness.

4.8 GROUND FAULT DETECTION IN DC SYSTEM

4.8.1 Ground Fault Detection in DC System (Off Line)

Remarks on ground fault detection in DC system may be appropriate here. DC system is usually ungrounded. For ground fault detection, two equal and high resistances (R) are connected between +ve and –ve terminals.

The center point is earthed through a Center Zero Ammeter (Figure 4.9). Voltage between positive pole and negative pole is 220V DC.

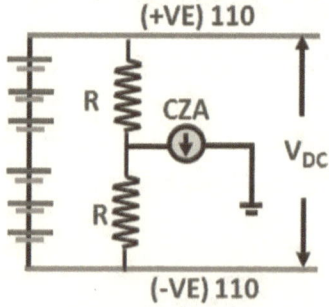

Figure 4.9: DC system

If fault occurs on positive pole, ammeter deflection is on one side of zero and if fault is on negative pole, the deflection is on other side of zero (Figure 4.10). Typical range of CZA is -100mA → 0 → +100mA. The value of resistor (R) is 3 to 4KΩ.

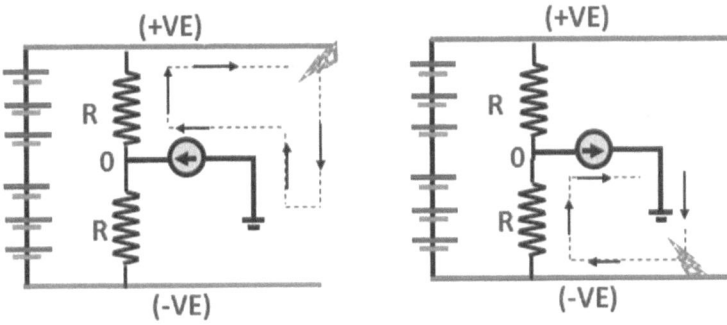

Figure 4.10: Ground Fault Current Flow

Fault on either positive or negative pole does not hamper the system operation. But to locate the ground fault in DC system is very tedious. A typical method to detect ground fault is shown in Figure 4.11. Each feeder is fed through a make before break switch. If CZA alarm comes, each feeder is switched from Normal (N) to Test (T) position. For the faulted feeder, CZA alarm on one battery vanishes and appears on other battery.

After the detection of first earth fault, it has to be located and cleared within a reasonable time (8 to 24 hours). Multiple DC earth faults hanging reflects poorly on O&M practices of the company.

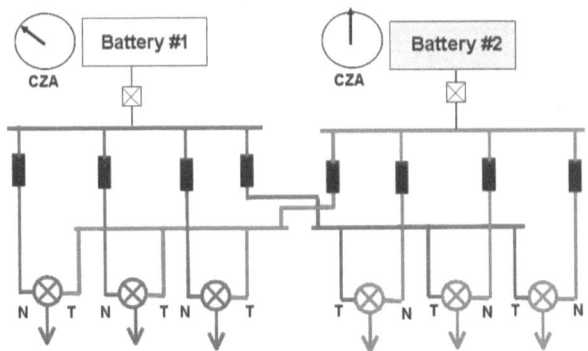

N - Normal position ; T - Test position

Figure 4.11: Ground fault detection in DC system

4.8.2 Ground Fault Detection in DC System (On Line)

To reduce time for detecting and locating earth fault in DC systems, on line systems are in vogue [33]. It consists of:
(a) Insulation monitoring device
(b) Earth fault sensors (CBCT)

a) Insulation monitoring device

It is electrically connected to DC system on both positive and negative pole (Figure 4.12). It is also connected to earth through protective conductor (PE) which offers return path for injected pulse. Controlled AC voltage pulse is superimposed on the DC system. The measuring cycle consists of positive and negative pulses of the same amplitude. The frequency of AC pulse is typically 8 to 10Hz and magnitude is 40V. Since the voltage pulse is applied simultaneously on positive and negative poles, it results in cancellation when the voltage is measured between positive and negative poles and DC voltage is unaffected. An insulation fault between system and earth results in flow of current to the measuring circuit. Insulation resistance value can be calculated with measured current against applied voltage pulse magnitude on positive and negative pole.

With insulation monitoring device, occurrence of DC earth fault can be detected. However, the location of earth fault cannot be detected. Thus, to locate on which feeder fault has occurred, CBCT is used.

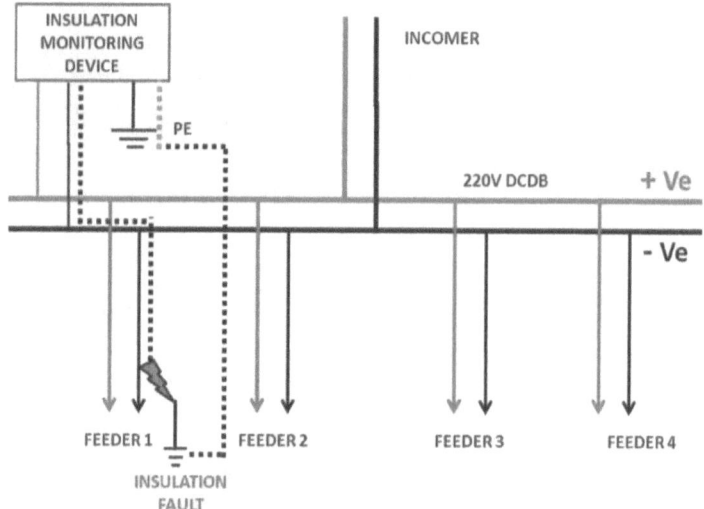

Figure 4.12

b) Earth fault sensors (CBCT)

Earth fault sensor is basically Core Balance Current Transformer (CBCT), which operates on the principle of net flux produced by current carrying conductors (single core unarmoured cables separate for +ve and –ve) that are made to pass through CBCT. The CBCT is of Spilt type or ring type. CBCT communicates with Insulation Monitoring Device using conventional metallic cables. Selection of CBCT is done based on type of cables, minimum fault current through main conductor to be detected and transformer ratio required for receiver device. The CBCTs are installed at each feeder compartment emanating from the DC distribution board. Connection diagram is shown in Figure 4.13.

If armoured cable is used, with the presence of ground fault, when AC pulse is applied, there will be a flow of current both in main conductor as well as armour. But the direction of current in main conductor is opposite to that in armour resulting in cancellation of flux. Net output of CBCT is zero. Hence this scheme will not work if bonded armoured cable is used for DC distribution.

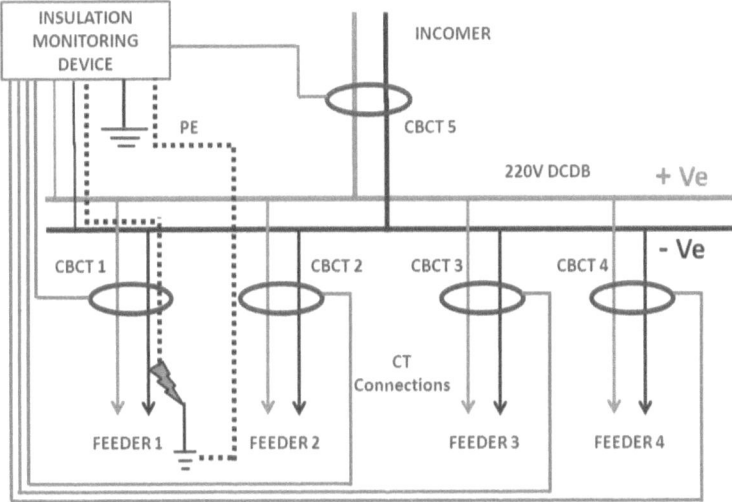

Figure 4.13

4.9 GROUND FAULT DETECTION IN AC SYSTEM

The most preferred method to detect ground faults in ungrounded system is through open delta PT connection shown in Figure 4.14.

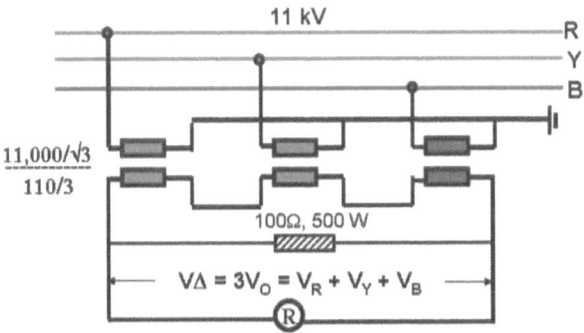

Figure 4.14: Open Delta PT Connection

For a line to ground fault, on 11kV system, from Eqn. (4.2B),
$$V\Delta = V_R + V_Y + V_B = \sqrt{3} \times 11kV$$

Voltage on secondary side of PT:

$$V_{REL} = \left[\frac{(110/3)}{(11000/\sqrt{3})}\right] \times \sqrt{3} \times 11000$$
$$= 110\ V$$

The reason for choosing secondary side PT voltage as (110/3) instead of usual (110/√3) becomes evident now. Relay rated for 110 V can be connected across open delta. If the secondary side PT rating is (110/√3), voltage across the relay will be √3 x 110 V. Relay rated for 110 V can be damaged in this case.

At some sites, over heating of PT is observed. It is probably due to resonance between magnetizing reactance of PT and connected system capacitance. This is termed in literature as 'ferro-resonance'. To avoid ferro-resonance, it is recommended to provide a damping resistor across the relay branch as shown in Figure 4.14. The resistor value is typically 100Ω. The resistor converts ungrounded system into very high resistance grounded system. Further discussion on this aspect will be covered in Sec 5.11.3.

Sometimes, auxiliary PT is used in open delta connection as shown in Figure 4.15.

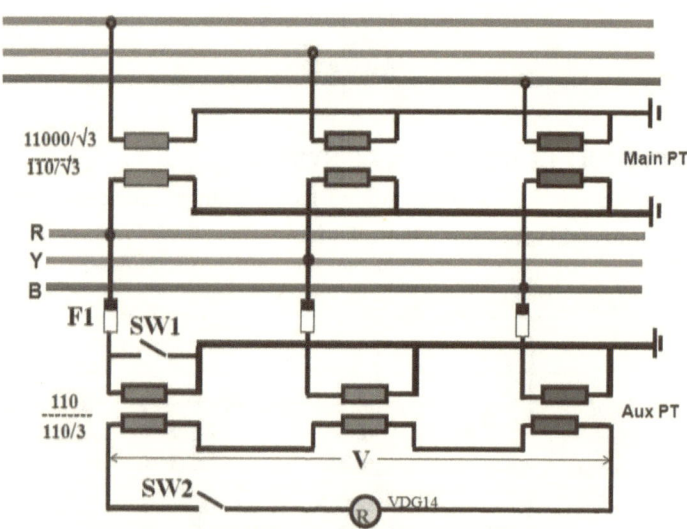

Figure 4.15: Auxiliary PT Connection

Switches SW1 and SW2 are used only when testing the scheme. Many times relay operation is found to be sluggish during testing. One of the reasons could be that the

primary winding of phase under test is *not shorted*. It is not enough to just remove the fuse. The correct test procedure is as follows:
  (i) Keep switches SW1 and SW2 off
  (ii) Measure open delta voltage 'V'. It should be nearly zero
  (iii) Remove fuse F1 on R phase
  (iv) Close switch SW1 (to simulate $V_R = 0$)
  (v) Measure open delta voltage 'V'. It should be nearly 110 Volts.
  (vi) Close switch SW2
  (vii) Check voltage relay operation
  (viii) Repeat the above procedure for Y phase and B phase

4.10 GROUND FAULT DETECTION ON SHIPS

The alternators supplying power in ships are in majority of cases rated for 440V. On land, LT system is solidly grounded as per IE regulation (refer Sec 5.8.3). But on ships, the system is ungrounded to ensure continuity of supply under single earth fault conditions. To detect earth faults, arrangement shown in Figure 4.16 is used. It is called as 'Three Lamp Method' for ground fault detection. Each lamp shall be rated for 440V.

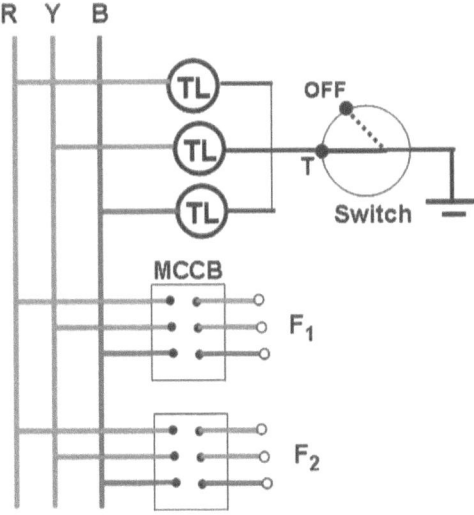

Figure 4.16: Earth indications (Healthy system)

Under healthy conditions all the three Test Lamps (TL) will glow dull uniformly as the voltage across each lamp is 240V (440/√3). Under earth fault condition on, say phase R, Test Lamp on R phase will be dimmer (dark) but the other two lamps will be bright (Figure 4.17). Voltage across R phase lamp is low (nearly zero) but the voltage across other two lamps will be 440V. The earth indication switch is normally kept in OFF position. The operator periodically puts the switch in Test position (T) to test for earth fault. In case of earth fault, the current returns through ship's hull which acts as the 'earth'. If earth fault indication comes, the operator has to open feeder by feeder to isolate faulted feeder.

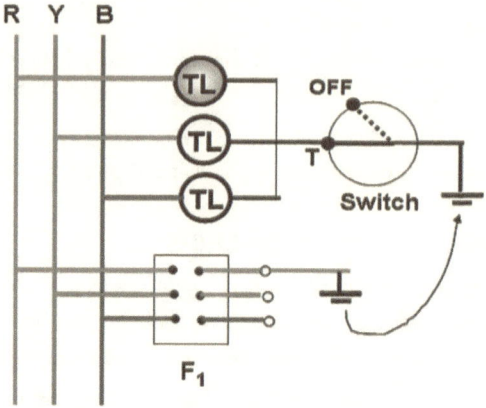

Figure 4.17: Earth indication (Faulted system)

4.11 SINGLE PHASE PT CONNECTION FOR GROUND FAULT DETECTION

A single phase PT is connected across R phase. Under and over voltage relays are connected as shown in Figure 4.18. For ground fault on R phase, under voltage relay picks up. For ground fault on Y phase or B phase, over voltage relay picks up as in these cases R phase voltage rises to $\sqrt{3}$ times normal value.

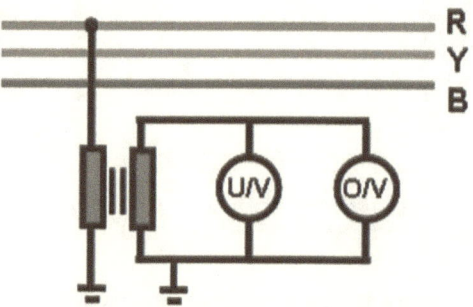

Figure 4.18: Single phase PT Connection

4.12 NEUTRAL INVERSION AND FERRO-RESONANCE

The single phase PT connection discussed in previous section, under certain system conditions, can lead to destructive events like ferro-resonance and inversion of neutral. For this reason, this connection is rarely used in practice now. However the analysis is included here as it is intellectually stimulating. The connection diagram is shown in in Figure 4.19. Source neutral is ungrounded. Single phase PT is connected between R phase to ground.

V_R, V_Y, V_B: Phase to Neutral impressed voltages
V_{NG}: Neutral to ground voltage
X_M: Magnetizing reactance of PT
X_C: System capacitance per phase

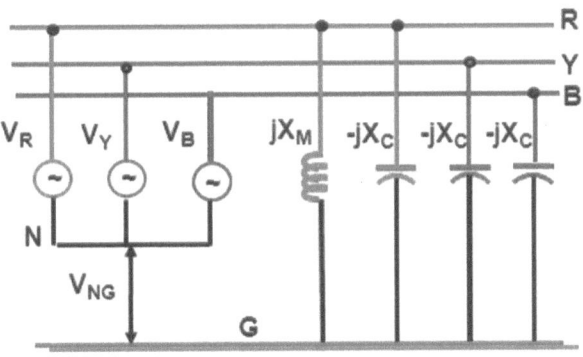

Figure 4.19

In Figure 4.20, Z is the equivalent impedance of X_C and X_M in parallel.

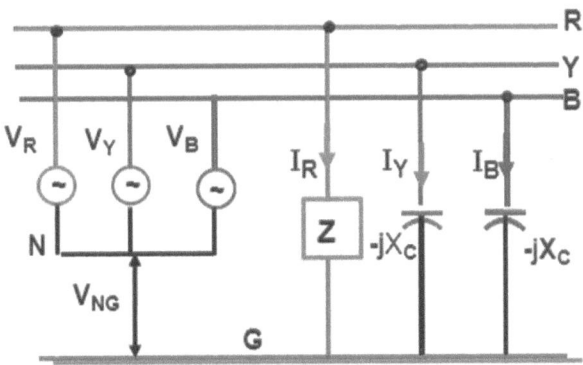

Figure 4.20

From Figure 4.20,

$$Z = \frac{(jX_M)(-jX_C)}{j(X_M - X_C)}$$

$$= \frac{-jX_M X_C}{(X_M - X_C)} \tag{4.3}$$

$$V_{RY} = I_R Z - I_Y(-jX_C) \tag{4.4}$$

$$V_{BR} = I_B(-jX_C) - I_R Z \tag{4.5}$$

From Eqn. (4.4) and (4.5)

$$V_{RY} - V_{BR} = 2I_R Z + jX_C(I_Y + I_B) \tag{4.6}$$

$$V_{RY} - V_{BR} = V_L \angle 0 - V_L \angle 120^0 = \sqrt{3}V_L = 3V_P \tag{4.7}$$

Since system is ungrounded

$$I_R + I_Y + I_B = 0$$

$$I_Y + I_B = -I_R \tag{4.8}$$

From Eqn. (4.6) to (4.8)

$$3V_P = 2I_R Z - j I_R X_C$$

$$I_R = 3V_P \frac{1}{(2Z - jX_C)}$$

R Phase voltage to ground

$$V_{RG} = I_R Z$$

$$= 3V_P \frac{Z}{(2Z - jX_C)}$$

$$= 3V_P \frac{1}{\left(2 - j\frac{X_C}{Z}\right)}$$

From Eqn. (4.3),

$$V_{RG} = 3V_P \frac{1}{\left[2 - jX_C \frac{(X_M - X_C)}{-jX_M X_C}\right]}$$

$$= 3V_P \frac{1}{3 - \left(\frac{X_C}{X_M}\right)} \tag{4.9}$$

Neutral to ground voltage

$$V_{NG} = V_P - V_{RG}$$

$$= V_P - 3V_P \frac{1}{3 - \left(\frac{X_C}{X_M}\right)}$$

$$= V_P \frac{1}{1 - 3\left(\frac{X_M}{X_C}\right)} \tag{4.10}$$

With V_P = 1 pu, the variation of Neutral to Ground Voltage V_{NG} with respect to (X_M/X_C) is shown in Figure 4.21

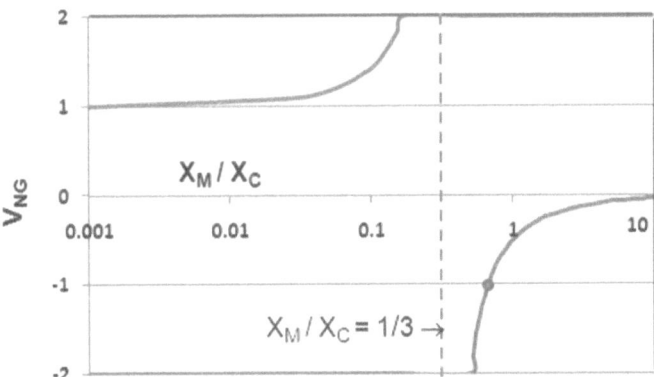

Figure 4.21

If the capacitive reactance is very much less than magnetizing reactance of PT $\{(X_M/X_C)$ ratio is large\}, neutral to ground voltage is nearly zero. In the absence of single phase PT, $(X_M/X_C) = \infty$, and $V_{NG} = 0$. This pertains to balanced condition.

When $(X_M/X_C) = 1/3$, $V_{NG} = \infty$

This condition, also known as *Ferro-resonance* (resonance between iron path in PT and connected system capacitance), leads to destructive over voltages.

If the capacitive reactance is very much greater than magnetizing reactance of PT $\{(X_M/X_C)$ ratio is small\}, *magnitude* of neutral to ground voltage is greater than 1pu.

For a solid fault on R phase, (X_M/X_C) is zero and V_{NG} = 1pu. It is the same as obtained from Eqn (4.2).

The peculiarity of Figure 4.21 is that V_{NG} is either greater than 1pu or less than zero (negative). Further discussions on this topic will continue in the next section...

With V_P = 1 pu, the variation of Phase to Ground voltage V_{RG} with respect to (X_M/X_C) using Eqn (4.9) is shown in Figure 4.22

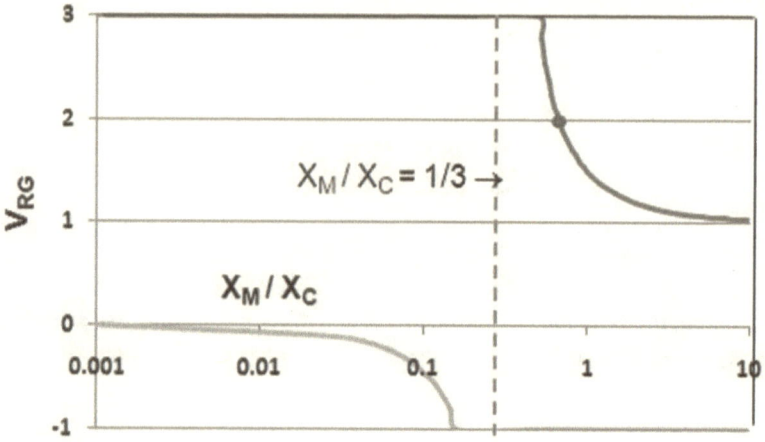

Figure 4.22

For a solid fault on R phase, (X_M/X_C) is zero and as expected $V_{RG} = 0$.

If the capacitive reactance is very much less than magnetizing reactance of PT $\{(X_M/X_C)$ ratio is large$\}$, phase to ground voltage is nearly one. In the absence of single phase PT, $(X_M/X_C) = \infty$, and $V_{RG} = 1$. This pertains to balanced condition.

When $(X_M/X_C) = 1/3$, $V_{RG} = \infty$

This is Ferro-resonance leading to destructive over voltages.

When $(X_M/X_C) = 2/3$, $V_{RG} = 2$pu. Corresponding neutral shift is -1pu (Figure 4.21).

The phasor diagram is shown in Figure 4.23. Under balanced condition, Neutral and ground are at the same potential and there is no 'Neutral Shift'.

When $(X_M/X_C) = 2/3$, Ground no longer remains within voltage triangle and this is called 'Neutral Inversion'. The same point is brought out in Sec 7.4.2 of [18].

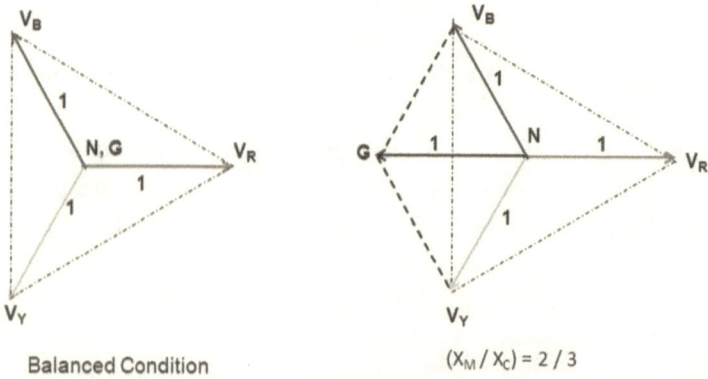

Figure 4.23

4.12.1 Similarity with Capacitor Switching

The peculiarity of the curve shown in Figure 4.21 is that V_{NG} does not lie between 0 to 1 pu; *either it is above 1 or below 0*. However this becomes obvious if we consider a simple (L-C) circuit as shown in Figure 4.24.

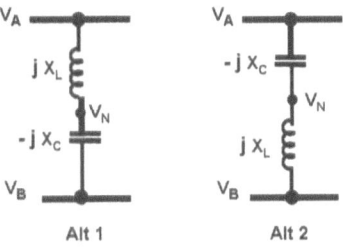

Figure 4.24: (L-C) Network

Let $V_A = 1$; $V_B = 0$

The intermediate voltage, from circuit analysis, for Alt 1:

$$V_N = 1 + \left[\frac{1}{\{(X_C/X_L) - 1\}} \right] \qquad (4.11)$$

If $(X_C/X_L) > 1$,
$V_N > 1$ i.e. $(V_N > V_A)$
If $(X_C/X_L) < 1$,
$V_N < 0$ i.e. $(V_N < V_B)$

For example, if $(X_C/X_L) = 5$, $V_N = +1.25$ and if $(X_C/X_L) = 0.2$, $V_N = -0.25$.
V_N is either above 1 or below 0.

Series reactors are provided with capacitors in power factor improvement circuits. The function of series reactor is to limit the inrush current when capacitor bank is switched on. Usually 0.5% to 6% reactors are employed in these schemes. Two alternatives of connecting the reactor are possible. In Alt 1, reactor is on bus side and capacitor is on neutral side. In Alt 2, capacitor is on bus side and reactor is on neutral side. Assume 6% reactor is used ($X_L = 0.06 X_C$).

For Alt 1, from Eqn(4.11), V_N is 106%. The maximum voltage to ground experienced by both the capacitor and reactor is 106%.

For Alt 2, the intermediate voltage is given by:

$$V_N = 1 + \left[\frac{1}{\{(X_L/X_C) - 1\}} \right] \qquad (4.12)$$

For Alt 2, from Eqn(4.12), V_N is -6%. The maximum voltage to ground experienced by capacitor is 100% and reactor is subjected to only a maximum of 6% *under steady state conditions*. Hence this is the preferred alternative for connecting the reactor. In a 11kV capacitor bank with 6% series reactor, steady state voltage across reactor is only 381V (0.06 x 11/√3kV). It may appear that LV reactor rated for 415V will suffice. But during switching, reactor is subjected to full voltage as brought out by simulation studies described in next section.

4.12.2 Capacitor Switching Simulation

33kV, 14.65MVAR capacitor bank with 1% reactor is considered for simulation. The network modeled in PSCAD is shown in Figure 4.25.

$V_P = 33/\sqrt{3} = 19$kV
$X_C = 33^2/14.65$
$\quad = 74.3345\,\Omega$

$C = 1/(2 \times \pi \times f \times X_C)$
$\quad = 42.8198\,\mu F$

$X_C/R = 500;\ R = 0.1487\,\Omega$
$X_L = 0.01 \times 74.3345$
$\quad = 0.7433\,\Omega$

$L = X_L/(2 \times \pi \times f)$
$\quad = 0.0024\,H$

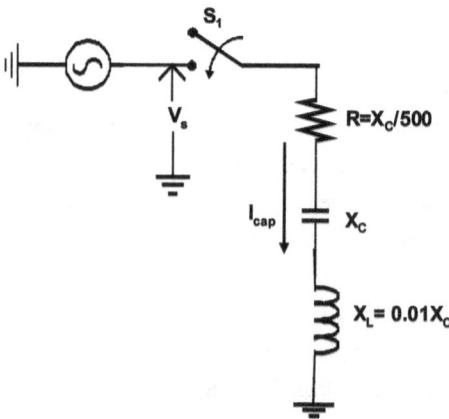

Figure 4.25: Capacitor with Small Inductor

Capacitor switching is simulated at t = 0.105 sec, when source voltage is passing through maximum. The source voltage (steady) and voltage at the reactor terminal (decaying) are shown in Figure 4.26.

During steady state, the voltage at reactor terminal is very small.

V_L = 0.01 x V_P = 0.19kV rms. However, during switching, the peak voltage across reactor may rise up to $\sqrt{2}$ x 19kV. Hence, the reactor must have the *same BIL* as that of the capacitor [13].

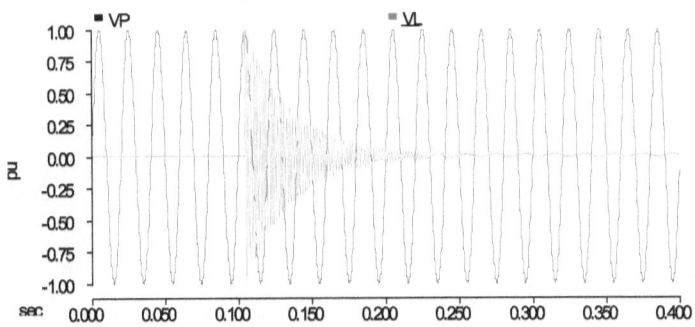

Figure 4.26: Reactor Terminal Voltage during Switching

Grounding of Electrical System – Grounded System

Chapter 5

5.0 INTRODUCTION

In Chapter 4, basic concepts of grounding were introduced. The essential features of ungrounded system were elaborated. In this Chapter we will progress to grounded systems [6]. The definition of 'effective grounding' is given. The features of Resistance grounding and Reactance grounding will be covered. The operation of Zig Zag connection will be demystified. Star-delta and open delta connections to ground the ungrounded system are explained. The relative advantages of each type of grounding are examined. The over voltage problems in non-solidly grounded systems are discussed. Next, corner earthing of delta tertiary is dealt with. At the end recommended CT secondary grounding practices are given.

5.1 REASONS FOR GROUNDING

It enables sufficient ground fault current to flow so that selective isolation of faulted section is feasible. During abnormal system conditions like fault, it minimizes the 'neutral shift' and limits the over voltages appearing on the system.

Ungrounded system is cursed with overvoltage problem. In grounded system the over current problem has to be solved.

5.2 GROUNDING LOCATIONS
   (i)   The neutral of star connected stator winding of generator, Figure 5.1.
   (ii) The neutral of star winding of power transformer, Figure 5.2.
   (iii) The neutral of 'grounding transformer', Figure 5.3.
   (iv) One corner of delta, Figure 5.4

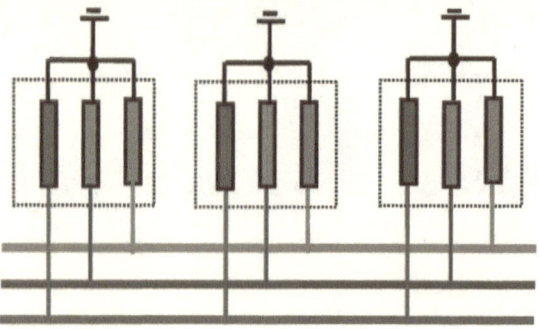

Figure 5.1: Neutral of star connected generators

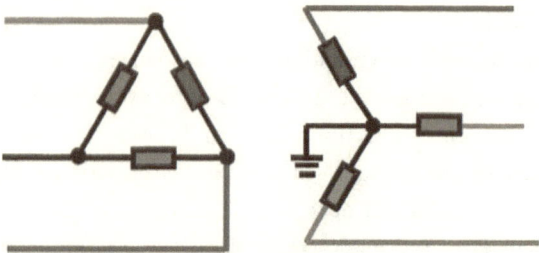

Figure 5.2: Neutral of star winding of transformer

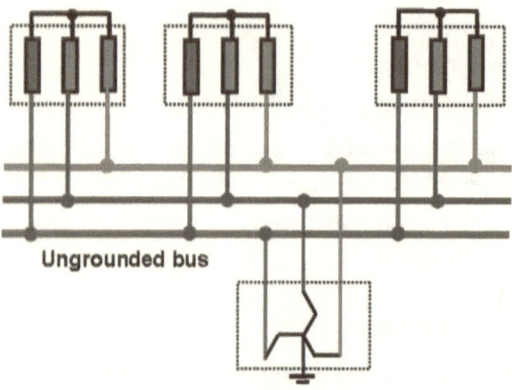

Figure 5.3: Neutral of grounding transformer

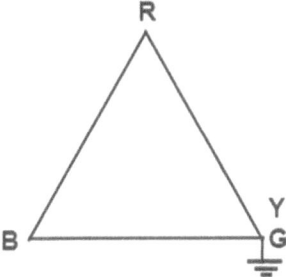

Figure 5.4: Corner earthed delta

5.3 GROUNDING METHODS
  (i) Solidly grounded system
  (ii) Resistance grounded system
  (iii) Reactance grounded system

The above classification is based only on the nature of external circuit connected between neutral and ground. Each of the above will be discussed in the sequel.

5.4 MEASURE OF GROUNDING EFFECTIVENESS
What constitutes 'effective' grounding is addressed by evaluating certain parameters. Based on extensive simulation studies, it has been established that if the parameters are within the specified range, the transient over voltages under disturbed system conditions are limited.

One such figure of merit [26] extensively used is given below (Cl 3.3 of IEEE Guide C62.92.4)

$$K_F = \frac{I_{1PH}}{I_{3PH}} \tag{5.1}$$

I_{1PH}: Single phase to ground fault current
I_{3PH}: Three phase to ground fault current

For effectively grounded system, $K_F > 0.6$.
Refer Table 5.1. Case1 qualifies as effectively grounded system. Other cases fall under non-effectively grounded system. Later it will be seen that cases 1,2 and 3 pertain to solidly grounded, low resistance grounded and high resistance grounded system respectively. A point worth emphasizing here is that *even if a small impedance (resistance or reactance) is introduced between neutral and ground, the system tends to become non-effectively grounded.*

Case	I_{3PH} (kA)	I_{1PH} (kA)	K_F
1	40	40	1
2	40	1	0.025
3	40	0.01	0.00025

Table 5.1

Another figure of merit is the 'Earth fault factor' - E_F

$$E_F = \frac{\text{Max phase to earth voltage of sound phases during ground fault condition}}{\text{Rated phase to earth voltage under healthy condition}}$$

For effectively grounded system, $E_F < 1.4$
For ungrounded system, $E_F = \sqrt{3}$
For solidly grounded system, $E_F \cong 1$

The standards (e.g. Cl 1.2.1 of IEEE Std 142) give a more precise definition for effectively grounded system [28].

"A system or portion of the system can be said to be effectively grounded when for all points in the system or specified portion thereof

$$\left(\frac{X_0}{X_1}\right) < 3 \text{ and } \left(\frac{R_0}{X_1}\right) < 1 \quad (5.2)$$

Typical values of X/R ratio for different elements are given below:
   Generator – 60 to 100
   Transformer – 10 to 15
   EHV Line – 5 to 15

Hence for solidly grounded system, R_0 is small and the second relation

$(R_0/X_1<1)$ is usually satisfied. As for the first relation, consider a typical EHV network shown in Figure 5.5.

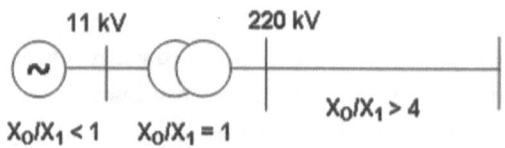

Figure 5.5: Typical EHV System

Let us now consider first relation ($X_0/X_1 < 3$)
For generator, X_0 (zero sequence reactance) is about 10% and X_1 (transient reactance) is about 25%. Hence $\left(\dfrac{X_0}{X_1}\right) < 3$.

For transformer, X_0 and X_1 are nearly same. Thus $\left(\dfrac{X_0}{X_1}\right) < 3$.

However, in case of EHV transmission line, $\left(\dfrac{X_0}{X_1}\right) > 4$.

Thus for the same system, at generator and transformer terminals, the system is effectively grounded. But at the end of a long radial EHV line, the system may not be effectively grounded.

5.5 REASON FOR CHOOSING $K_F \geq 0.6$

Ignoring resistance,
Positive Sequence Impedance = X_1
Negative Sequence Impedance = $X_2 = X_1$
Zero Sequence Impedance = X_0

For 3 phase fault,

$$I_{3PH} = \dfrac{1}{X_1}$$

For 1 phase fault,

$$I_0 = \dfrac{1}{X_1 + X_2 + X_0}$$
$$= \dfrac{1}{2X_1 + X_0}$$

$$I_{1PH} = 3I_0 = \dfrac{3}{2X_1 + X_0}$$

$$K_F = \dfrac{I_{1PH}}{I_{3PH}} = \dfrac{3X_1}{2X_1 + X_0} = \dfrac{3}{2 + \left(\dfrac{X_0}{X_1}\right)}$$

From Eqn. (5.2) at the limit $\dfrac{X_0}{X_1} = 3$

$$\dfrac{I_{1P}}{I_{3P}} = 0.6$$

5.6 SOLIDLY GROUNDED SYSTEM

The neutral is connected to ground without any explicit external element like resistor or reactor. This is a subset of 'Effectively Grounded System', in which $K_F = 1$.

5.7 ADVANTAGES OF SOLIDLY GROUNDED SYSTEM

(i) The substantial flow of ground fault current enables accurate detection and location of ground faults. The Earth Fault Factor E_F is nearly 1 and hence transient overvoltage is minimum. Also, the neutral shift during ground fault is markedly less (almost one third) compared to ungrounded system (Refer Sec. 4.4).

The ground fault relay (51N) connected in residual circuit offers sensitive protection (Figure 5.6) for feeders. A separate CBCT is not required for ground fault detection. Since ground fault currents are high, sensitivity is not an issue and higher setting of (say 80%) is recommended especially if CT ratio is lower. Further discussions on this topic will be covered in future Chapter on 'Relay Setting and Coordination'.

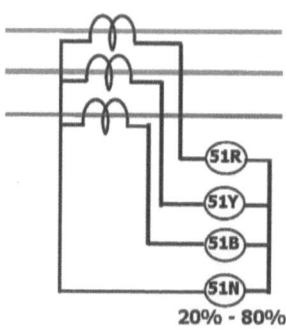

Figure 5.6: Ground relay Residual connection

(ii) Lightning arrestor rated for 80% can be used. For example, in a 132kV system LA rated for 106kV will suffice.

(iii) The over voltage factor for PT can be 1.5 pu (instead of 1.9 pu). Refer Cl 5.3 of IEC 60044-2 [27].

(iv) Earthed grade cables can be used (Refer Sec 5.13)

5.8 DISADVANTAGES OF SOLIDLY GROUNDED SYSTEM

5.8.1 Zero Sequence Current Circulation

It permits flow of zero sequence currents. Third harmonic and multiples of third harmonic currents are zero sequence currents. Every generator produces certain (minimum) amount of third harmonic voltage.

If the neutrals of generators on a common bus are solidly grounded, substantial third harmonic currents can circulate between generators resulting in increased heating. In this case, neutral of only one generator is grounded and neutrals of other generators are kept ungrounded (Refer Sec 7.2 and 7.6).

5.8.2 Stator Core Damage

In case of solidly grounded system, the ground fault current magnitude is high. The core damage at the point of fault in rotating machines like generator and motor will be high. To limit the damage to the core, manufacturers allow only a limited ground fault current. This information is usually provided in 'core damage curves' supplied by manufacturer. Core damage curves from two manufacturers are shown in Figure 5.7 and 5.8. In Figure 5.7, for example, ground fault current up to 25A is tolerated for 1 sec. This curve is used as a guide when selecting NGR and setting stator earth fault relays in generator protection.

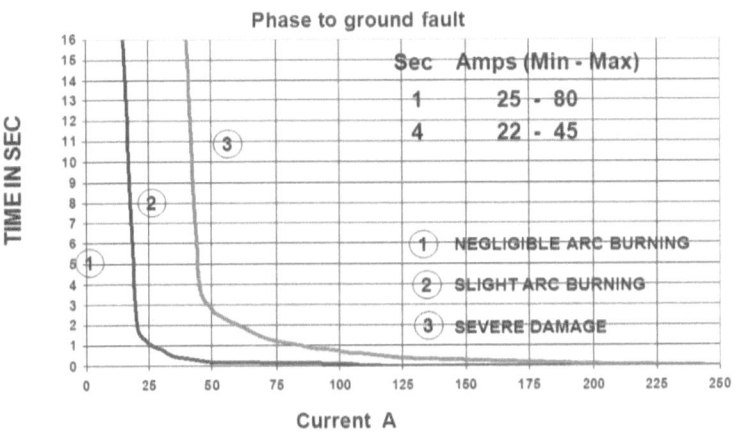

Figure 5.7: Generator – core damage curve (Vendor A)

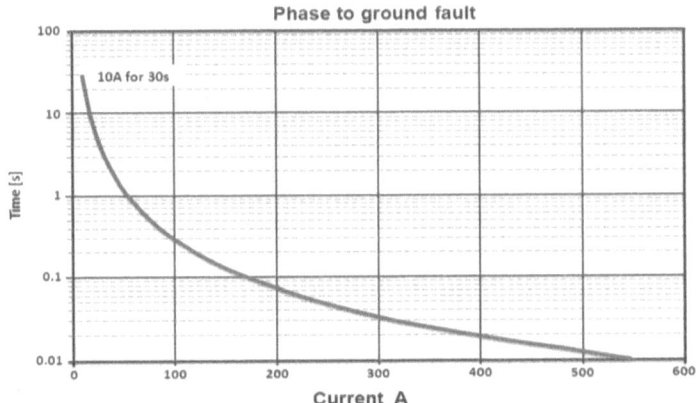

Figure 5.8: Generator – core damage curve (Vendor B)

Typically ground fault currents are limited to 10 to 15A in case of large generators. In old generators asphalt based insulations were used for generator stator coils and are more prone to develop earth faults (in generators stators that have core lengths of over 3 meters and have non-base load operation) particularly in the end stamping area where there is a greater likelihood of tape separation and girth cracking of the insulation.

In modern machines using epoxy based insulations, earth faults largely result from looseness of stator bars (due to shrinkage or slot discharge activity) that typically fail at the slot ends or extensions due electromagnetic forces that originate in the end-winding, and sometimes in the slot regions due to wear or impact of the insulation.

The extent of damage is generally proportional to the amount of energy that is dumped into the particular location of core at the time of the fault. Considering 100 A of fault current flowing for one second, the amount of material that would melt for a large 500MW generator would be to the extent of 1 Kg. This is quite a substantial amount of material and such damage would generally necessitate core re-stacking. Considering, the extent of damage it would be preferable to restrict the current to within 10A to 15A.

Decisions for core repair (partial, grinding or re-stack) are based on tests that are performed on the core. Two tests are generally used. ELCID tests are performed at a low flux levels to identify shorted stampings or flux loop tests at higher flux levels. If shorts are found then the repair resorted to is grinding the faulty area and repeating ELCID in the faulty area.

If the area is still faulty then perform Electrochemical Etching and retest the faulty area.

If the fault persists then partial restacking will be resorted to (i.e. only few stampings will be removed and the same replaced after varnishing and staggering) if the fault is at the slot end.

Winding damages in rotating machines are not of serious concern. The repairs can be done at site by local rewinding agency with the help of OEM (Original Equipment Manufacturer). If the fault is within the core, it will have to be completely re-stacked. In most cases this cannot be carried out at site. The machine has to be sent back to manufacturer's works for repair. The economic loss due to non availability of generator for prolonged period is huge.

With regard to core monitoring systems, there are essentially two types: one which has general temperature detection, i.e. by using RTDs located in direct contact with the core at the bottom of the slot. With this method it would not be possible to detect localized developing stator core faults.

The other is with the help of Gas Analysis systems for machines which are Hydrogen cooled. In this case, a kind of DGA is done of the gases that could evolve as a result of high stator core temperatures, to relate the gas concentrations from insulation pyrolysis with the temperature of the developing defect. When an earth fault occurs, the method will not be quick enough to protect the machine, unless the earth fault has been caused by a developing core damage that has been gradually leading to increase in local core temperatures.

In summary, it is essential to limit the energy discharged into the slot to a minimum. One tool is to limit the current magnitude and the other is very fast clearance of fault. Core damage curve gives the relationship between the two.

5.8.3 Type of Grounding at Different Voltage Levels

Since rotating machines are not present in voltage levels from 22kV and above, usually these systems are solidly grounded. At EHV level solid grounding is universally adopted for two reasons: (a) cost of insulation at EHV level is high (b) primary protections clear the fault within 5 cycles.

If rotating machines are present at 3.3kV, 6.6kV and 11kV levels, the systems are grounded through resistor or reactor to limit the ground fault current. If rotating machines are not present at these voltage levels, the systems can be solidly grounded.

In case of LT (415V) system, even if rotating machines are present, the system is solidly grounded to conform to Cl 61-(1)(e) of IE rules [22]. Since LT system is also handled by 'general public', for safety reasons solid grounding is mandated. Sufficient ground fault current is allowed to flow so that protective devices can operate and clear the faults at the earliest. Of course, core damage at the point of fault in rotating machines will be high.

If large number of rotating machines (up to 175 kW) is present at LT level, it may be worth considering resistance grounded system even at this level to limit the ground fault current. LT buses can be segregated into those supplying rotating machines with resistance grounding and those supplying static loads like lighting and heaters with solid grounding. Refer Figure 5.9. The scheme shall be implemented in a controlled environment like power plant or industrial plant manned by professionals. Special application shall be made to the local electrical inspector who will review and approve the scheme in these cases.

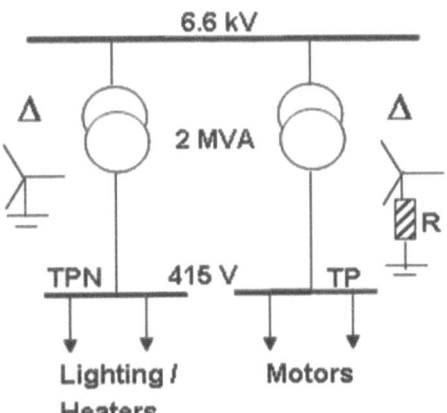

Figure 5.9: LT distribution

5.9 RESISTANCE GROUNDED SYSTEM

A resistor is connected between the neutral and ground. The reasons for limiting the ground fault current are as follows:

(i) In rotating machines, winding damage is tolerable but core damage is not.
(ii) Reduced burning and melting in electrical equipment
(iii) Reduced mechanical stresses (F α I^2) compared to solidly grounded system
(iv) Possibility of restrike/arcing faults very less compared to ungrounded system

Depending on the value of limiting fault current, it is further classified as high resistance grounding and low resistance grounding.

5.9.1 High Resistance Grounded System

In High Resistance Grounded system (Figure 5.10), the ground fault current (I_F) is limited to within 10A to 15A. The value of resistor is selected such that for a ground fault, current through resistor I_R is equal to (or greater than) total system capacitive current I_C. The phasor diagram is shown in Figure 5.11.

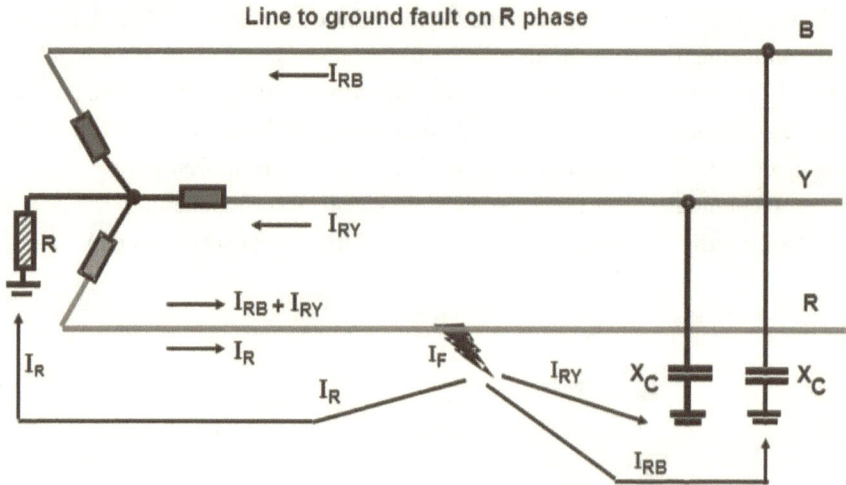

Figure 5.10: Resistance Grounded System

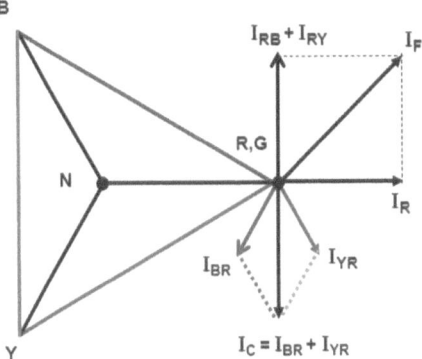

Figure 5.11: Phasor Diagram

Following are achieved by inserting high resistance in neutral:
(i) In case of ungrounded system, fault current is I_C (Figure 4.5). In case of resistance grounded system fault current is $I_F = \sqrt{2}I_C$. Though magnitude of I_F is greater compared to I_C, power factor of I_F is 0.7 whilst that of I_C is 0. It implies during arcing faults when current is interrupted at natural zero the voltage is maximum in ungrounded system while it is less (70%) in high resistance grounded system. Since voltage available in the ionization path is less, there is less chance of restrike in resistance grounded system.
(ii) The resistor provides the path for capacitance to discharge rapidly as time constant (CR) is in msec. This prevents voltage build up and reduces possibility of high transient over voltages during arcing earth faults.

Consider a 11kV system. Let the ground fault current be limited to 10A. The value of NGR(Neutral Grounding Resistor) is approximately given by:

$$R_G \cong \frac{\left(\frac{11000}{\sqrt{3}}\right)}{10}$$
$$= 635\,\Omega$$

5.9.1.1 Neutral Grounding Transformer (NGT)

One method for achieving the above is to connect a 635 Ω resistor directly in the neutral circuit. But a more economical solution is to connect the resistor across the NGT. This uses the elementary fact that an impedance Z connected to the secondary side of transformer gets reflected as $T_R^2 Z$ on primary side where T_R is the turns ratio. Refer Figure 5.12

for illustration. Instead of connecting 1000Ω resistor on primary side, 10Ω resistor can be connected on secondary side. The primary current (I_P) drawn in both alternatives is same (0.3A).

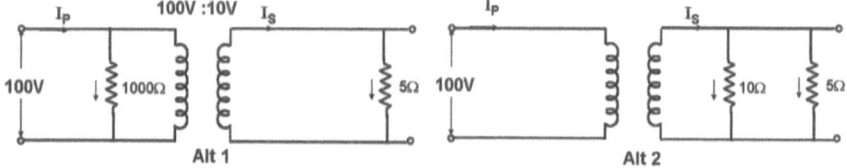

Figure 5.12: Concept of reflected impedance

The scheme with NGT is shown in Figure 5.13. Consider 600MW, 667MVA, 20kV generator. Ground fault current is to be limited to 5A. Value of resistance is worked out as follows.

Phase Voltage = $20/\sqrt{3}$ = 11.55kV
Rated Current = $667/(\sqrt{3} \times 20)$ = 19.26kA
Required resistance R_G = 11.55/5 = 2.31kΩ
Consider NGT ratio as 16/0.24kV
Turns Ratio = 16/0.24 = 66.7
Value of resistor on LV side R'_G = $2310/66.7^2$ = 0.52Ω

By connecting a 0.52Ω resistor on NGT secondary side, for a line to ground fault on generator terminal ground fault current is limited to 5A whilst the rated current of generator is 19,260A. This is the power of impedance transformation. More details are given in Sec 7.8.

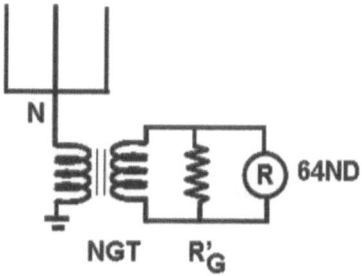

Figure 5.13: Resistor connected to NGT

The use of low resistance low voltage resistor results in economical design. A voltage relay (Neutral Displacement Relay) is connected across the resistor to detect ground faults.

The majority of application of high resistance grounded system is in neutral grounding of large generator units to limit ground fault current to a small value so that core damage is minimum. Since ground fault current magnitude is very low, only voltage based schemes are used to detect ground faults. Related subject of stator earth fault protection (95% and 100%) will be covered separately.

5.9.2 Low Resistance Grounded System

In low resistance grounded system, the ground fault current is limited to about 400A. Another widely used criterion is to limit the fault current to some percentage (say 50%) of rated current of source transformer or generator. Low resistance grounding is widely adopted in MV systems (3.3kV, 6.6kV and 11kV) where rotating machines (induction and synchronous) are present. This is a compromise between high resistance grounded system and solidly grounded system. Compared to high resistance grounded system, the core damage at the faulted location will be more but the machine sizes are much smaller in this case. The ground fault current is neither too high (in kA) nor too low (less than 15A). Since sufficient ground fault current exists, current based protection using CBCT is widely used for both detection and isolation of faulted feeder.

Also the motors are present at the tail end of distribution systems (Figure 5.14). The relays (R_1) on motor feeders do not require coordination with upstream relays (R_2 & R_3) and faults within motor are cleared by R_1 within 50 to 100msec.

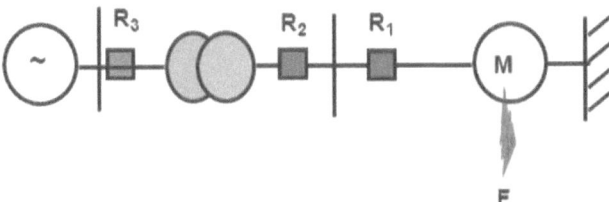

Figure 5.14

On 11kV system, with ground fault current limited to 400A, value of NGR is approximately given by:

$$R_G \cong \frac{\left(11000 / \sqrt{3}\right)}{400} = 16\,\Omega$$

The value of resistor is small (2310Ω in case of high resistance grounded system described in previous section). The resistor is directly connected between neutral and ground (Figure 5.15).

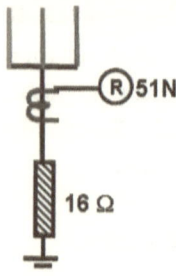

Figure 5.15: Low Resistance Grounded system

Since the ground fault current is limited to within 400A, heat dissipation (I^2R loss) is not excessive and enclosures can be designed so that temperature rise is within limits. Resistor elements made up of stainless steel or cast iron are commonly used. NGR on support insulators connected to neutral of LV (11kV) side of transformer is shown in Figure 5.16. The enclosure with louvers is mounted on a pedestal approximately 2.5M above ground so that it is normally not approachable. The dimensions of enclosure are 700x700x1500 mm. The enclosure is compact since it is designed only for 10sec duty. Special requirements of enclosure earthing were covered in Sec 3.4.7.

Figure 5.16:

Since sufficient ground fault current exists, current based protection is feasible. Current relay in neutral circuit of source transformer is provided as back up to feeder faults and to protect the NGR. Ground relay connected in residual circuit, as in Figure 5.6, can be used for feeder protection. The ground fault relay in the range of 10% to 40% is adequate. CBCT is usually provided on feeder circuit for positive pickup.

5.9.3 Comparison Between Ungrounded and Grounded Systems

Parameter	Ungrounded	High Resistance Grounded	Low Resistance Grounded	Solidly Grounded
$I_{1\text{-}\Phi}/I_{3\text{-}\Phi}$	< 0.5%	< 1%	5% to 20%	> 60%
Transient overvoltage	3 to 6 pu	Not more than 1.5 to 2.5 pu		
Arrestor rating		100%		80%
PT voltage factor		190% for 8 hours		150% for 30 Sec
Fault location	No	Perhaps		Yes
Immediate disconnection after ground fault	No	Optional		Yes
Expected repair (winding) after ground fault		New winding insulation		
Expected repair (core) after ground fault	NIL		Perhaps Core Stacking	Core Stacking
Multiple faults	Often		Seldom	

Table 5.2

5.10 REACTANCE GROUNDED SYSTEM

Here, a reactor is connected between neutral and ground. Two types are available in reactance grounding.

In the first type the ground fault current is almost nil and is termed as 'Resonant Grounding'.

In the second type ground fault current is substantially higher (minimum 60% of three phase fault current) and is termed as 'Effectively Grounded System'.

5.10.1 Resonant Grounding

The reactor is called 'Peterson coil'. The value of reactor is selected such that, for a ground fault, current through reactor I_X is equal to total system capacitive current I_C (Figure 5.17, Figure 5.18). From Eqn (4.2A),

$$I_C = I_{BR} + I_{YR} = \frac{3}{X_C}$$

$$I_X = \frac{1}{X_L}$$

$$\frac{1}{X_L} = \frac{3}{X_C}$$

$$X_L = \frac{X_C}{3}$$

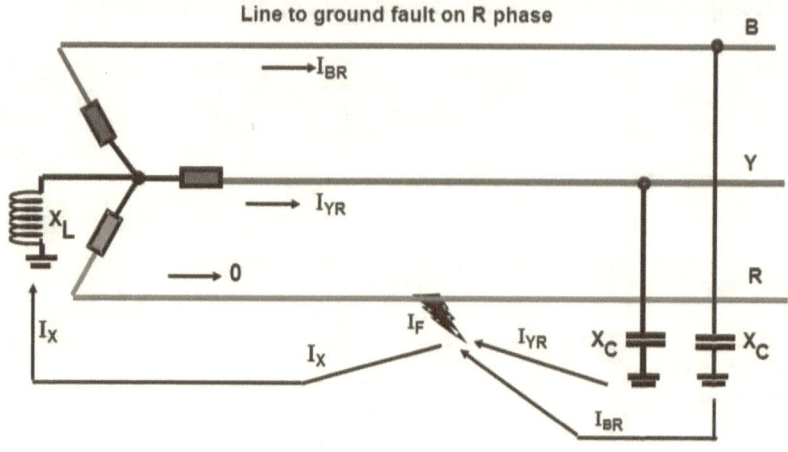

Figure 5.17: Reactance grounded system

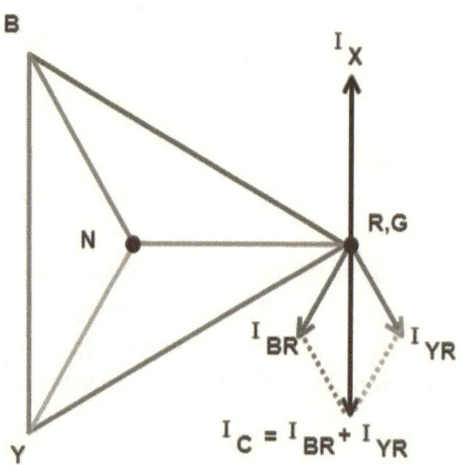

Figure 5.18: Phasor diagram

Consider a typical value of 2A per phase for charging current in 11kV system.

$$X_C = \frac{\left(11000/\sqrt{3}\right)}{2} = 3175 \, \Omega$$

$$X_L = \frac{3175}{3} = 1058 \, \Omega$$

K factor of reactor $(X_L/R) \cong 20$

5.10.1.1 Advantages of Resonant Grounded System
(i) The current through the reactor I_X almost nullifies capacitive current I_C in case of a ground fault. The fault current is practically zero. Hence the possibility of restrike is remote as the arc is self-extinguishing.
(ii) Substantial zero sequence voltage (V_0) is available for fault detection through open delta PT.

5.10.1.2 Disadvantages of Resonant Grounded System
(i) The system capacitance will change due to operational procedures when some feeder cables are taken in service or out of service. The reactor has to be tuned to match the system capacitance. This is a time consuming task in a running system. If the reactor taps are left untouched even when system capacitance has changed (due to network switching), the main purpose of 'resonant grounding' is not realized.
(ii) In rotating machines, core damage is perceptible only if the ground fault current exceeds 20A. Hence reducing the ground fault current to practically nil does not enhance further the core damage withstand performance.

Resonant grounding was tried in a few large units in the past (reactor instead of resistor in Figure 5.13). But in modern system design, reactance grounding is rarely used. Only resistance grounding (high or low) is used for neutral grounding of rotating machines. Further discussions on generator neutral grounding practices are given in Chapter 7.

5.10.2 Effectively Grounded System
Features of effectively grounded system were already introduced in Sec 5.4. Two relationships are defined in this context.

a) $K_F \geq 0.6$.
   K_F = (Single Phase to ground fault current) / (3 Phase Fault current)
   = I_{1P}/I_{3P}

   This is frequently used by field engineers as it is easy to understand and implement.
   For solidly grounded system, $K_F \geq 1.0$.
   For ungrounded system, $K_F \cong 0$

b) E_F (Earth Fault Factor) ≤ 1.4
   E_F = Maximum Line to ground voltage on healthy phase during fault / Rated Line to ground voltage

When a Fault occurs on Phase B of Feeder A (Figure 5.19), B phase Bus voltage dips and B phase voltages of all connected feeders (FDR B & C) also dip. Depending on type of source grounding, R and Y phase bus voltages will change. In case of ungrounded system they will rise to 1.732pu. In case of effectively grounded system, voltages of healthy phases will rise to a maximum of 1.4pu. Hence all equipment connected to FDR B and FDR C will also experience overvoltage on R phase and Y phase even though fault is on FDR A. The point to note is that fault on one feeder will affect voltages of *all connected feeders* on the bus until the faulted feeder is disconnected by protection.

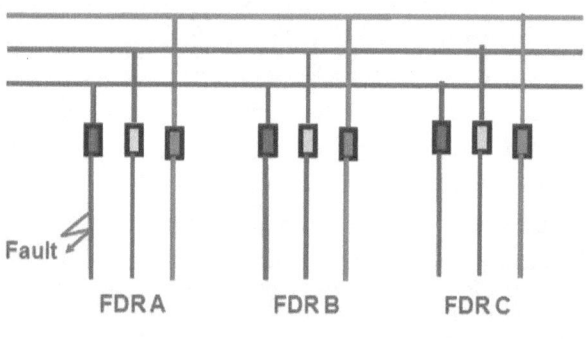

Figure 5.19

5.10.2.1 Illustration with Practical Example

The underlying concept is explained with a practical example of existing distribution system of a major metropolis [14].

Power supply to the city is derived through multiple voltage transformations. The bulk power is stepped down at Transmission Stations. A typical Transmission Station (T/S) has a number of 220/33kV, Star – Zig Zag transformers. The choice of vector groups of transformers at different voltage levels will be covered later in Chapter on 'Transformers'. The 220kV Star neutral is solidly grounded whilst 33kV Zig Zag neutral is grounded through NGR (Neutral Grounding Reactor). Each transformer feeds 5 to 6 Receiving Stations (Refer Figure 5.20).

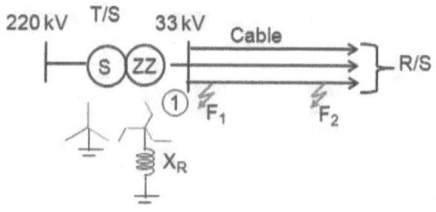

Figure 5.20

At the Receiving Station (R/S), step down transformer has the following rating: 33/11kV, 20MVA, Delta – Zig Zag. Secondary neutral is solidly grounded. Each transformer feeds 5 to 6 Ring Mains. Each Ring Main (Figure 5.22) serves 5 to 10 Sub-Stations. At each Sub-station, 11/0.44kV Distribution Transformers (DT) step down power and feed LT distribution system.

The vector group and type of grounding at Transmission, Receiving and Substation are given in Figure 5.21.

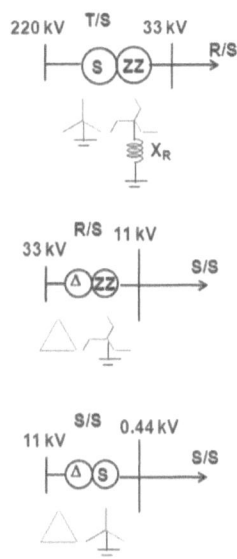

Figure 5.21

At Transmission Stations, secondary of transformer is 'effectively grounded'. At Receiving Stations and Sub-stations, secondary of transformer is 'solidly grounded'. The meaning of 'solidly grounded' is that there is no intentional intervening impedance present between the transformer neutral and ground. It may be noted that 'solidly grounded' system is a subset of 'effectively grounded system'. A 'solidly grounded' system is 'effectively grounded' but an 'effectively grounded' system need not be 'solidly grounded'.

Refer Figure 5.20 (T/S). Rating of transformer is 125MVA, 220/33kV,
  Star – Zig Zag,
  $Z_P = Z_N = 15\%; Z_0 = 2.5\%$.
  The value of Z_0 is low as the secondary winding is Zig Zag.
  Assume 220kV is infinite bus (source impedance is 0)
  Three phase fault current at 33kV Bus1 = I_{3P} = $(125/0.15)/(\sqrt{3} \times 33)$ = 14.6 kA
  Rated phase voltage = $33/\sqrt{3}$ = 19.05kV

The secondary neutral is earthed through NGR to limit the ground fault current to a desired value. Assume $X_R = 1\Omega$.

Consider a ground fault on Phase R very near to Bus1 (F_1 in Figure 5.20). Voltages and currents can be worked out using "symmetrical Component Analysis". From results of simulation,

$I_{1P} = I_R = 10.18$ kA
$K_F = 10.18/14.6 = 0.7$

Since $K_F > 0.6$, the system for this fault is effectively grounded.
This can be reconfirmed from voltage rise on healthy phases during fault.

$V_R = 0$
$V_Y = V_B = 22kV (116\%)$
$E_F = 1.16 < 1.4$

Case 2

The same example is repeated with ground fault (R Phase) on cable at 3.6 KM away from Bus1 (F_2 in Figure 5.20). The cable parameters used for simulation are given below [12].

Cable size: 3C x 400 mm² Al
$Z_P = Z_N = 0.08 + j\,0.117\ \Omega/KM$
$Z_O = 0.646 + j\,0.644\ \Omega/KM$
From results of simulation:
$I_{1P} = I_R = 5.88$ kA

The voltages at Bus1 for far end fault are given below:
$V_R = 9.39kV (49\%)$
$V_Y = 22.13kV (116\%)$
$V_B = 19.91kV (105\%)$
$E_F = 1.16 < 1.4$

This brings out an important fact that even though ground fault current is only 40% of three phase fault current at Bus1 (5.88/14.6 = 0.4), the voltage rise at the Bus1 is still within limits (<1.4 pu). Since the fault is at remote location, voltage of faulted phase at Transmission Station (V_R) is substantial. *Hence all other feeders connected to Bus1 do not experience over voltage. The deciding criterion to declare system is effectively grounded is E_F rather than K_F*

Case 3

Refer Figure 5.22 (R/S). Transformer rating is 20 MVA, 33/11kV, Delta – Zig Zag,
$Z_P = Z_N = 12.5\%$; $Z_0 = 3\%$.

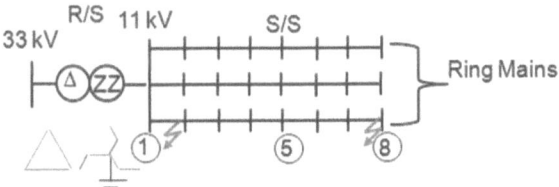

Figure 5.22

Three phase fault current at Bus1 = I_{3P} = (20/0.125)/($\sqrt{3}$x11) = 8.4 kA
Rated phase voltage = 11/$\sqrt{3}$ = 6.35kV
Secondary neutral is solidly grounded.

Case 3.1
Ground fault is simulated on Phase R very near to Bus1. From results of simulation,
$I_{1P} = I_R$ = 11.25 kA
K_F = 11.25/8.4 = 1.3

The reasons for K_F much greater than 1 are: (i) primary is delta connected (ii) secondary is solidly grounded and (iii) zero sequence impedance is much smaller (3%) as secondary is Zig Zag connected.

The phase voltages at Receiving Station bus 1 are:
V_R = 0
$V_Y = V_B$ = 5.6kV (88%)
E_F = 0.88 < 1.4

The system is effectively grounded for fault very near to Receiving Station bus. None of the connected feeders will experience over voltage.

Case 3.2
In Figure 5.22, a sample ring main is considered for simulation. It has 7 substations and distance between substations is 500 meters. The cable parameters used for simulation are given below:

Cable size: 3C x 300 mm² Al
$Z_P = Z_N$ = 0.123 + j 0.102 Ω/KM
Z_O = 1.173 + j 0.427 Ω/KM

For a ground fault on R phase at remote Bus 8,
$I_{1P} = I_R$ = 2.184 kA

The ground fault current is only 26% of three phase fault current at Bus1 (2.184/8.4 = 0.26),

The phase voltages at faulted Bus 8, intermediate Bus 5 and Receiving Station Bus 1 are shown in Table 5.3.

	Bus 8		Bus 5		Bus 1	
	Voltage kV	E_F	Voltage kV	E_F	Voltage kV	E_F
V_R	0		2.513	0.40	5.865	0.92
V_Y	8.134	1.28	6.961	1.10	5.933	0.93
V_B	9.951	1.57	8.409	1.32	6.380	1.00

Table 5.3

(i) At the remote Bus 8, E_F > 1.4, hence locally it is 'non-effectively grounded'.
(ii) At intermediate Bus 5, E_F is marginally less than 1.4, just managing to be categorized as 'effectively grounded'.
(iii) At the Receiving Station Bus 1, E_F is ≤ 1 and it is 'effectively grounded'.

From results of above case studies, the following observations are made:

(a) Irrespective of fault location, E_F at Receiving Station is ≤ 1.0. This has important implication that at Receiving Station, voltages of un-faulted phases *do not rise* above normal phase voltage. Hence voltage of other feeders (Ring Mains) connected to the bus will not experience over voltage.

(b) As the fault location is moved away from Receiving Station, E_F at remote location is higher. It can cross the threshold limit of 1.4. At the remote locations it is no longer effectively grounded system. But in substations closer to Receiving Station even on the faulted feeder, E_F < 1.4. Thus over voltage is limited to *local area near to faulted point far away from Receiving Station.*

The standards recognize this fact. Even in case of solidly grounded system, some parts of system may not be effectively grounded for particular fault location. The same point was brought out in discussions on Figure 5.5. *The aim of solid grounding is to limit over voltages to local areas and over voltages are not felt globally over entire system for fault in any one location.*

In this context, the relevant extract from IEEE Guide C62.92.4, Cl 3.3 [26] is reproduced below:

"The overvoltage on un-faulted phases is also of concern because it is applied to the equipment of customers served from distribution transformers connected from phase to neutral on four-wire systems. Thus, even if arrester application is not a limiting factor, the

E_F must not be allowed to increase to a level that can impose intolerable over voltages on customer equipment.

As a rule of thumb, E_F at the substation should not exceed 1.25, which is obtained approximately when $X0/X1 = 2$. Preferably E_F should not exceed 1.1, which requires an $X0/X1$ of 1.3 or less. *At locations remote from the substation, the E_F will exceed these values because of the effects of line impedance. However, the lower values at the substation are desirable to mitigate the effect of the line impedance and to localize the over voltages near the fault location rather than requiring the whole system to withstand them.* It is realized however, that higher $X0/X1$ ratios have been used satisfactorily".

It is possible to choose NGR value so that $K_F = 0.4$ to 0.5, with E_F nearly equal to 1.4 for faults very near to source transformer, anticipating lower ground fault current. But in this case, no margin is available in E_F. For any fault even slightly away from transformer, voltage at local substation will rise resulting in $E_F > 1.4$. This is the reason why the standards recommend that for effectively grounded system, NGR is sized such that $K_F \geq 0.6$.

Summarising:
 (i) Size NGR based on $K_F \geq 0.6$ for a ground fault on terminal of transformer
 (ii) Grounding effectiveness at remote locations is based on evaluating E_F at these locations
 (iii) Irrespective of type of grounding, use 100% arrestor for voltages 33kV and below.

More than 70% faults are single phase to earth faults. It is important to positively identify and isolate these faults. Current based earth fault protections are more sensitive and selective than voltage based system.

In solidly grounded system high magnitude of earth fault current is always ensured for faults anywhere in the system. It is easy to design sensitive earth fault detection system. However the damage at fault point could be severe. Also equipment (Transformers, cables, cable joints) which experiences the let through current, undergoes higher dynamic stress. If we restrict the earth fault current below a certain level by introducing impedance in the neutral, the healthy phase voltages rise to L-L values thereby stressing the insulation of *all* equipment connected to the system. This is also detrimental to the health of the equipment particularly in a network with aging equipment.

Effectively earthed system is balance between the two. We get sufficiently large current ensuring positive relay operation; at the same time the healthy phase voltages do not rise to dangerous levels.

In effectively earthed system, ground fault current magnitude is limited but still in kA. Hence only a reactor with high X/R ratio can be used in neutral. Usually the reactor is sized for 10 second duty. Resistor can't be used as the current is very high (more than 10kA in the above example) and size of resistor and associated enclosure to dissipate heat (I²R loss) will be very high and impractical to manufacture. The picture of reactor connected to neutral of 33kV side of transformer (X_R Figure 5.20) is shown in Figure 5.23.

It is cylindrical with 600 mm dia and 800 mm height. Size wise, it is similar to Neutral Grounding Resistor shown in Figure 5.16. Reactor is sized for 8kA for 10 sec. Resistor is sized for 400A for 10sec.

Figure 5.23

5.11 GROUNDING THE BUS

To establish grounding in ungrounded system, the two widely used methods are Zig Zag grounding and star-delta grounding. Open delta grounding is a sub set of star-delta grounding.

5.11.1 Zig Zag Grounding Transformer

5.11.1.1 At Balance in Transformer

The basic principle of conventional transformer operation is the Ampere Turns balance.

If expressed in per unit, the current (both in magnitude and angle) must be equal in primary and secondary windings on same limb of transformer. Any difference between primary and secondary current is used only in setting up the flux and is called exciting current or no load current. If transformer is energized with secondary open, the current flows only on primary and it is the exciting current or magnetizing current.

5.11.1.2 Winding Current During Normal Condition

The connection diagram is shown in Figure 5.24. The peculiarity in this case is that on the same limb of transformer, one winding (called Zig) carries current from one phase and the other winding (called Zag) carries current from other phase. We can consider Zig and Zag similar to primary and secondary windings. Under charged condition let the terminal currents be $1\angle 0°$, $1\angle -120°$ and $1\angle -240°$. Since currents in Zig and Zag windings are *not* same both in magnitude and angle, these can be only exciting current to set up the flux in the core. The flux is proportional to net current in Zig and Zag winding.

In Limb 1, the net current = $1\angle 0° - 1\angle -120° = \sqrt{3}I\angle 30°$

In Limb 2, the net current = $1\angle -120° - 1\angle -240° = \sqrt{3}I\angle -90°$

In Limb 3, the net current = $1\angle -240° - 1\angle 0° = \sqrt{3}I\angle 150°$

Thus the only +ve (or –ve) sequence currents the grounding transformer can carry is exciting current which is very small (less than 1A).

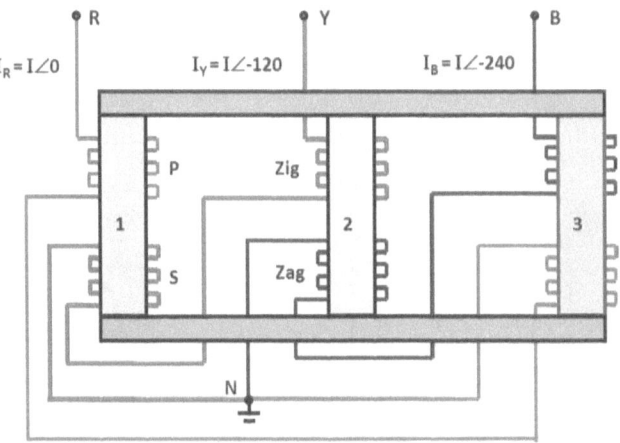

Figure 5.24

Except during ground fault (explained in next section) which is for very short duration of less than a second, the winding currents are negligible during its entire life. Hence, other than from core loss there is no heating internally from winding. The constant exchange of moisture between paper and oil that happens in normal transformer due to flow of load current is absent in Zig Zag NGT.

Though NGT is a small passive device compared to power transformer, it is essential to perform DGA at periodic intervals and monitor the quality of almost stagnant oil. If required furon analysis may also be required in some cases to check healthiness of paper insulation. Since core loss is the only significant loss in NGT, CRGO of good grade (M4 or better) silicon steel laminations shall be used to minimize constant losses.

5.11.1.3 Winding Current During Ground Fault Condition

During a ground fault, the fault current must return back to NGT neutral. Calculation of fault currents using theory of Symmetrical Components will be covered in a later Chapter. At this juncture, it is enough to point out that ground fault current is $3I_0$. I_0 is the zero sequence current flowing in all three phases whose magnitude and phase angles are same. The scene when zero sequence currents are flowing is shown in Figure 5.25. The current in Zig winding is, say $1\angle 0°$ and the current in zag winding is also $1\angle 0°$. This conforms to the fundamental principle of transformer theory stated above in which both primary and secondary currents are equal in magnitude and phase angle.

The transformer offers impedance to the flow of zero sequence current corresponding to conventional leakage impedance. Thus, this connection facilitates the flow of ground fault current ($3I_0$) in ungrounded system.

The ground fault current magnitude can be limited by inserting resistor or reactor in NGT neutral as shown.

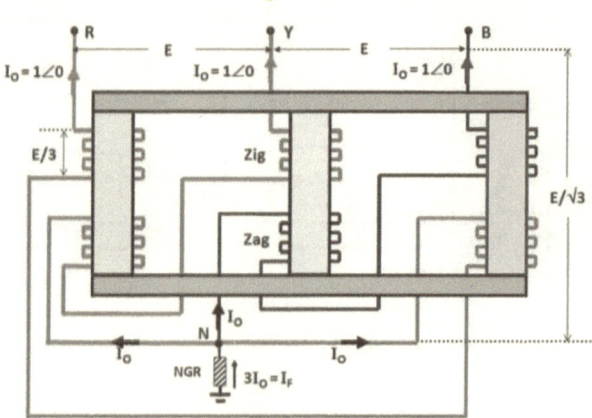

Figure 5.25

5.11.1.4 Phasor Diagram

The phasor diagram is shown in Figure 5.26.

Line voltage = E

Voltage across each winding = E/3

Voltage to neutral = $E/\sqrt{3}$

Current in each winding = $I_0 = I_F/3$

Rating of grounding transformer in MVA = $\sqrt{3} \, E \, I_0 = (E/\sqrt{3}) \, I_F$

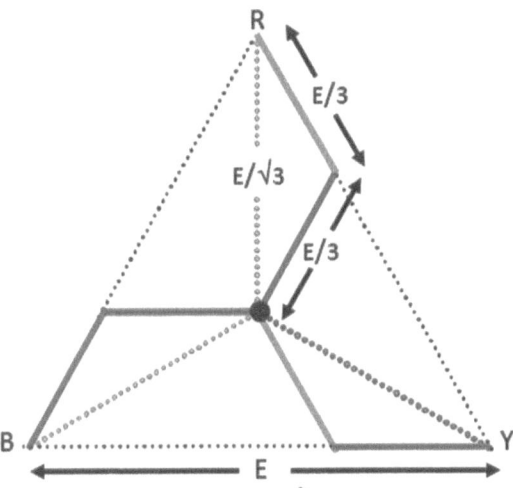

Figure 5.26:

The zero sequence leakage reactance (Z_0) is expressed on the above base MVA. In conventional transformer the leakage reactance is about 10%. But for grounding transformer, the leakage reactance can be specified even up to 100%. This is possible as it does not carry load current under normal conditions and regulation is not an issue. The desired ground fault current can be achieved by specifying Z_0 of NGT and NGR (if required). Both NGT and NGR are rated generally for 10 seconds. Hence its cost is very less (10% to 15%) compared to a conventional transformer of same rating.

5.11.1.5 Current Distribution for Ground Fault

Typical current distribution for ground fault is shown in Figure 5.27.

From system studies, Ground fault current $I_F = 3I_0 = 9kA$
Zero sequence Current $I_0 = 3kA$
Turns Ratio $TR = (220/\sqrt{3})/33 = 3.85$
Current on star side = I_0/TR
= 0.78kA

A line to ground fault on delta side of transformer gets reflected as line to line fault on star side of power transformer in presence of NGT.

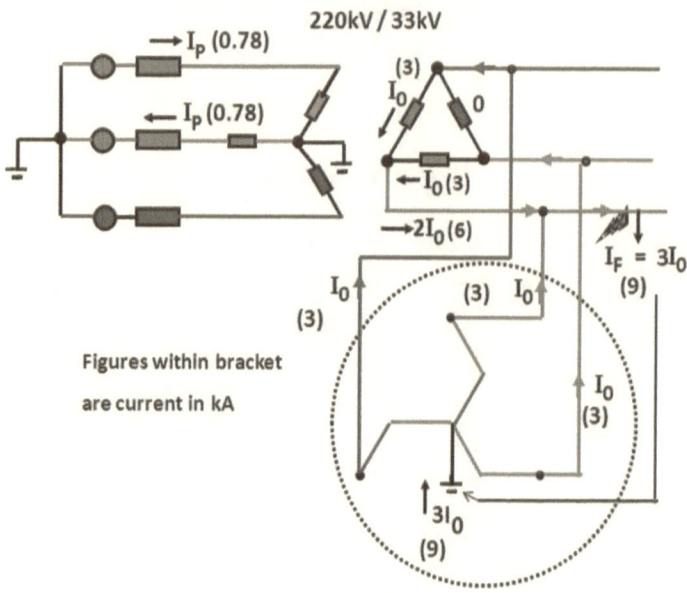

Figure 5.27

5.11.1.6 Miscellaneous topics on Zig Zag Grounding Transformer

It is very important to specify grounding transformer parameters correctly without ambiguity. Otherwise the equipment offered by vendor may not meet actual user requirement. In Appendix 5-1, resolution of this problem is discussed with numerical examples.

Another issue, though not very common, is choice of grounding when ground fault current is to be limited to a value which is neither too low nor too high. This is dealt in Appendix 5-2.

Sizing of NGT is discussed in Appendix 5-3.

Sizing of NGR is covered in Appendix 5-4.

In EHV switchyards in remote areas, NGT is sometimes provided with Auxiliary winding to derive LV supply. Appendix 5-5 deals with problems in supplying single phase loads.

In Appendix 5-6, short circuit testing methods of Zig Zag NGTs are covered.

5.11.2 Star-Delta Transformer

A conventional star-delta transformer can be used for grounding the bus (Figure 5.28).

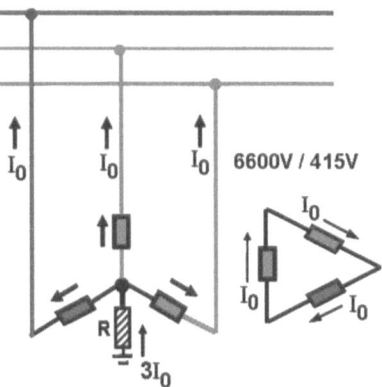

Figure 5.28: Star – Delta transformer

Though Zig Zag transformer will be preferred due to its lower cost, star-delta transformer can be used if a spare one is available. Under normal conditions, the transformer draws only the exciting current which is less than 1%. But for ground fault current it offers low impedance corresponding to zero sequence leakage reactance. I_0 flows in star winding and equivalent I_0 circulates in delta satisfying the Ampere-turn balance requirement for transformer operation.

This connection (star – delta) needs to be employed with caution as sometimes it leads to 'inadvertent grounding' as explained in Sec 6.12A and Sec 6.15A.

The ground fault current can be limited to any desired value by providing the resistor. The resistor can be connected in two ways – connected in primary neutral circuit or delta winding.

5.11.2.1 Resistor in Primary Neutral

Let the rated voltage of bus be 6.6kV and the resistor value be 250Ω. Refer Figure, 5.28.

Ground fault current $I_F = 3I_0$

$$= \frac{\left(6600/\sqrt{3}\right)}{250}$$
$$= 15.3 \text{ A}$$
$$I_0 = 5.1 \text{A}$$

5.11.2.2 Resistor on Delta Winding

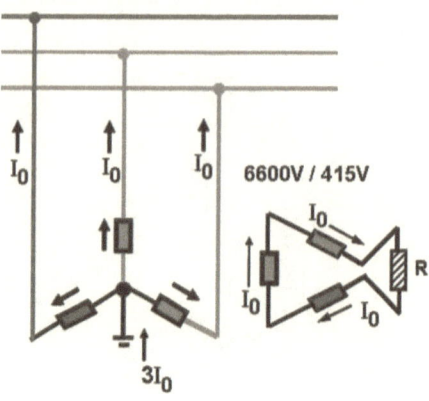

Figure 5.29: Resistor on delta side

Refer Figure 5.29. The required I_0 on star side is 5.1A.

$$\text{Turns Ratio TR} = \frac{\left(6600/\sqrt{3}\right)}{415}$$

$$= 9.18$$

I_0 on delta side = 5.1 x 9.18
$$= 46.8 \text{ A}$$

Resistance on delta side = $\frac{(3 \times 415)}{46.8}$ = 26.6 Ω

In general, $Rs = \frac{(R_D \times TR^2)}{9}$ (5.3)

R_S: Resistor value on star neutral in ohms
R_D: Resistor value on delta side in ohms

TR: Turns Ratio = $\frac{V_{STAR}^{PHASE}}{V_{DELTA}^{PHASE}}$

Even though the primary neutral is solidly grounded, the system behaves like resistance grounded system because of the presence of resistor on the delta side. The star winding in this case has to be rated for full line voltage.

5.11.3 Open Delta PT Grounding

Conceptually it is same as star – delta grounding. Open delta PT is used to detect ground faults in ungrounded system as described in Sec 4.9. A Ferro-resonance damping resistor, (typically 100Ω, 300W) is connected across the relay. Refer Figure 5.30.

From Eqn (5.3),

$R_D = 100 \Omega$

$$TR = \frac{\left(6600/\sqrt{3}\right)}{\left(110/3\right)}$$

$= 103.9$

$$R_S = \frac{\left(R_D \times TR^2\right)}{9}$$

$= 120 \text{ k}\Omega$

$$I_F = \frac{\left(6600/\sqrt{3}\right)}{R_S}$$

$= 32 \text{ mA}$

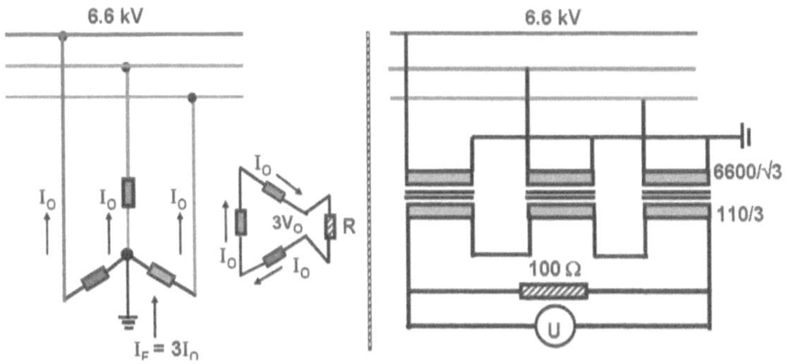

Figure 5.30: Open Delta PT with Resistor

An ideal ungrounded system is converted to a very high resistance grounded system with fault current limited to a very small value.

5.12 VARIATION OF VOLTAGE WITH FAULT CURRENT

Depending on type of grounding, the phase and line voltages change under ground fault condition. Line voltage triangle is isosceles for solidly grounded system and equilateral for ungrounded system. Refer Figure 5.31.

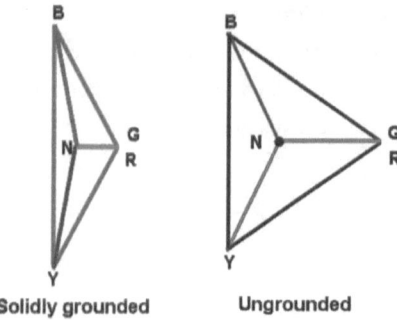

Solidly grounded   Ungrounded

Figure 5.31: Phasor Diagram

Under faulted condition, phase voltage and line voltage for unfaulted phases are equal. For fault on phase R,

$V_{YG} = V_{YR}; V_{BG} = V_{BR}$

By varying the value of NGR from zero to a very high value, conditions corresponding to solidly grounded, low and high resistance grounded and ungrounded systems can be simulated. The results of the simulation studies are shown in Figure 5.32. The striking feature is that line and phase voltages of unfaulted phase remain almost equal to $\sqrt{3}$pu until the fault current reaches a high value corresponding to solidly grounded system. Only when the ground fault current reaches around 8500A corresponding to solidly grounding, the line and phase voltages drop down to 1pu. The open delta voltage ($V\Delta = V_R + V_Y + V_B$) also exhibits a similar trend.

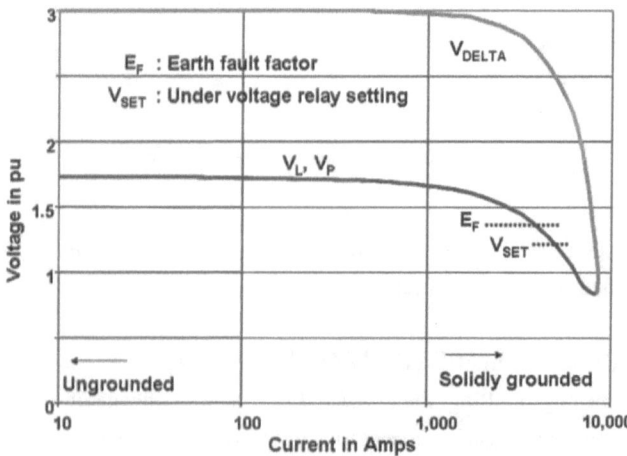

Figure 5.32: Variation of voltage with fault current

The following observations can be made:
(i) The voltage to ground of unfaulted phases remains at almost √3pu except for solidly grounded system. Consider a non-solidly grounded system. Let a ground fault occur on one of the cables from the switchgear. Not only the faulted cable but all the other cables connected to the switchgear will also experience overvoltage until the fault is cleared.

The connected equipment on the cable network like transformer and motor also experience high voltage during ground faults. Thus there is a cumulative stress on insulation of all the equipment after a ground fault.

(ii) Under-voltage relays are used for protection purpose. Assume the under-voltage relay is connected *across the line* and set at 70%. The line voltages remain almost at √3pu except for solidly grounded system (Figure 5.32). The relay will not pick up until the line voltage falls below say 70% of √3pu, i.e., V_{SET} = 1.2 pu. Thus line connected under voltage relays may not operate if used in non-solidly grounded system during ground faults. The preferred solution is to connect under voltage relays between phase and ground. This confirms a well known fact that the most reliable handle to detect voltage unbalance is the phase voltage and not the line voltage or the open delta voltage.

5.13 UNEARTHED GRADE (UE)/EARTHED GRADE (E) CABLE SELECTION

Some clarification is required regarding specification on grade of cable to be used depending on type of system grounding. As per Cl 2.5 of IS 7098-2 [24], definition of earthing system from cable insulation point of view is given below:

Earthed system - An electric system which fulfils any of the following conditions:
  (a) *Neutral-point or the mid-point connection is earthed in such a manner that, even under fault conditions, the maximum voltage that can occur between any conductor and the earth does not exceed 80 percent of the nominal system voltage;*
  (b) *The neutral-point or the mid-point connection is not earthed but a protective device is installed which automatically cuts out any part of the system which accidently becomes earthed; or*
  (c) *In case of ac systems only, the neutral-point is earthed through an arc suppression coil with arrangement for isolation within 1h of occurrence of the fault for the non-radial field cables and within 8h for radial cables, provided that the total of such periods in a year does not exceed 125h.*

Unearthed system – An electric system which does not fulfill the requirement of the Earthed system.

The above definition is also in line with Sec 4.1 of IEC 60502-2 [31].

Since in practical power systems all earth faults are cleared automatically within 1 second by protective relays irrespective of type of grounding (high resistance, low resistance, effectively grounded, solidly grounded), point (b) above is satisfied and *earthed cable can be specified In all cases*. In addition point (a) is also satisfied in case of effectively grounded system

($E_F < 1.4$, i.e., 80% of $\sqrt{3}V_P \cong 1.4\ V_p$). In case of solidly grounded system, $E_F \cong 1$.

Also the cables are designed to withstand much higher voltages. As per IS 7098-2, Cl 19.7.1 and Cl 19.7.2, following are specified:
  (i) Type test/Acceptance Test: Cable shall withstand without breakdown $3U_O$ applied between conductor and armour/screen for a period of *4 hours*.
  (ii) Routine Test: Cable shall withstand without breakdown specified voltage between conductor and armour/screen for a period of *5 minutes*.

For earthed grade cable, acceptance values are given in Table 5.4. These values are much higher than the value to which cable will be subjected to during faults. In conclusion, earthed grade cable (category A as per Cl 4.1of IEC 60502-2) can be specified *irrespective of type of system grounding*.

Voltage Grade U_O/U in kV	Type Test Voltage In kV for 4 hours	Routine Test Voltage In kV for 5 minutes
1.9/3.3	5.7	10
3.8/6.6	11.4	12
6.35/11	19.1	17
12.7/22	38.1	32
19/33	57	48

Table 5.4

5.14 PHASE VOLTAGE AND ZERO SEQUENCE VOLTAGE DURING GROUND FAULT

5.14.1 Phase Voltage
Phase voltage is high at source and almost zero at the fault point. Under voltage relay located near the fault location responds.

5.14.2 Zero Sequence Voltage
Source (generator) does not intentionally produce any zero sequence voltage and hence zero sequence voltage at source is nearly zero. At the point of ground fault, phase voltage at

faulted point collapses but zero sequence voltage is high. Refer Figure 5.33. Under voltage relay connected to phase PT and over voltage relay connected to open delta PT respond. Thus in both cases, voltage relays close to fault only respond.

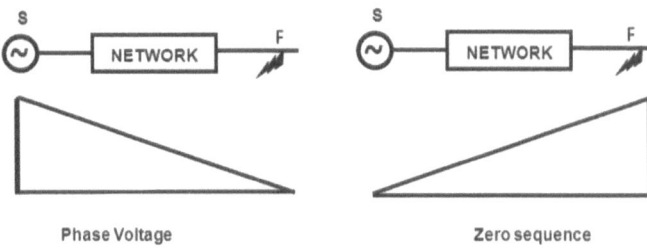

Figure 5.33

Also to be noted is that the phase voltage at faulted point is nearly zero *irrespective* of type of grounding of source. However zero sequence voltage at faulted point varies widely depending on type of grounding. It is high in ungrounded system and low in solidly grounded system. For illustration, zero sequence voltage V_0 is evaluated at the faulted point F_2, Figure 5.20. Values for three types of source grounding obtained from simulation are given below:

Ungrounded source, V_0 = 19kV

Effectively grounded source (X_R = 1Ω), V_0 = 12.3kV

Solidly grounded source = V_0 = 10.5kV

It is myth to assume that neutral shift does not occur in solidly grounded system, only its magnitude is less.

5.15 EARTHING OF DELTA

In power engineering, one of the 'mysteries' is delta connection. Some intriguing features of delta connection are given in [2].

Another aspect which has not been widely covered is 'earthing' of delta which appears as oxymoron as delta is inherently 'unearthed' system. There are advantages and disadvantages of earthing delta. There are two methods to earth a delta – center tap earthed delta and corner earthed delta. Salient points of these two methods are given below.

5.15.1 Ungrounded Delta

We recapitulate the features described in Sec 4.3 & 4.4. In the following discussions, 'Line voltage' refers to Line to Line voltage, e.g., V_{RY}. 'Phase voltage' refers to Line to ground voltage, e.g. V_{RG}.

In case of ungrounded delta, the line voltage is triangle is unaffected. However no 'reference' ground is available. Hence the phase voltage is 'floating' one. Depending on stray capacitance to ground for each phase, voltage to ground of each phase can be different. Refer Figure 5.34.

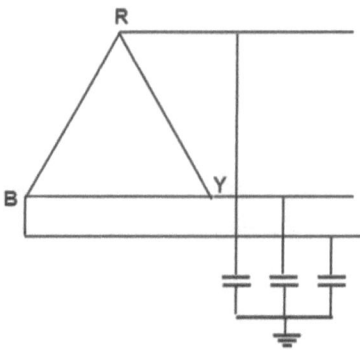

Figure 5.34: Ungrounded System Balanced Condition

In Figure 5.35, for balanced case with all stray capacitances equal, the three phase voltages are same and equal to $V_L/\sqrt{3}$.

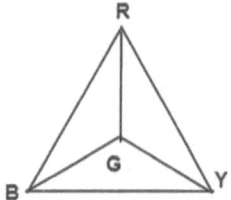

Figure 5.35: Balanced Condition

In Figure 5.36, for unbalanced stray capacitances, the voltages to ground of the three phases are different. However the line voltage triangle is unchanged. All the phase voltages are still less than line voltage.

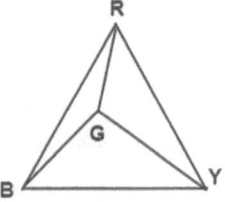

Figure 5.36: Unbalanced Condition

In Figure 5.37, situation is similar to 'neutral inversion' case. This case might arise if a single phase PTs is connected to ground on one phase and its exciting current is comparable to stray capacitance current. In this case, one of the phase voltages can be even higher than line voltage. This was discussed in depth in Sec 4.12.

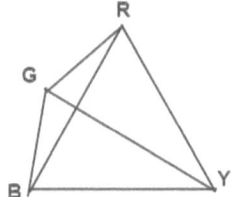

Figure 5.37: Neutral Inversion

In summary, in case of ungrounded delta, line voltages are rock steady but phase voltages are not well defined. Phase voltages could be less than, equal to or even greater than line voltages. In this case, phase to ground clearance should be same as that for phase to phase clearance.

The above situation is very similar to ungrounded DC system. The voltage between Positive Pole and Negative Pole will be as per requirement. However, voltage to ground of Positive Pole and Negative Pole could be very different.

5.15.2 Center Tap Earthing

To 'stabilize' the delta, one winding (YB in Figure 5.38) is earthed at its center tap. It is not 'correct' to label the center point of the winding as 'Neutral'. The connection to the center tap is usually referred as 'grounded conductor' rather than 'neutral conductor'. Since the grounding is on YB winding, R phase is termed as 'Red leg' or 'High leg'.

Following voltage relationships are applicable:

Line voltages: $V_{RY} = V_{YB} = V_{BR} = V$

Phase Voltages (not in strict sense): $V_{YG} = V_{BG} = V/2$; $V_{RG} = (\sqrt{3}/2) V$

This connection is archaic and legacy from delta – delta (centre tap earthed) connection used in the past. It is hardly used in practice now and is only of academic interest.

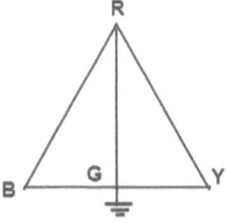

Figure 5.38: Central Tap Earthed

5.15.3 Corner Earthed Delta

Another method to 'stabilize' the delta is to earth one corner of delta. Refer Figure 5.39. In this case also, line voltage triangle is unaffected. However, one of the phase voltages (V_{YG}) is zero as it is 'earthed'. In this case, all line voltages (V_{RY}, V_{YB}, V_{BR}) and two phase voltages (V_{RG} and V_{BG}) are equal. There is no ambiguity in phase voltages.

One disadvantage is that the voltage to ground of ungrounded phase is 'always' equal to line voltage. Hence it is mandatory that phase to ground clearance should be same as that for phase to phase clearance.

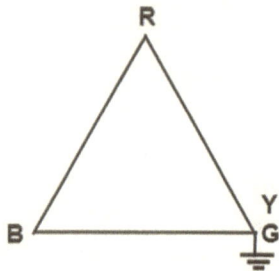

Figure 5.39: Corner Earthed Delta

Application of corner earthing in practice is limited to earthing of unloaded delta tertiary of large ICTs. For example consider 400/220/33kV autotransformer with 33kV tertiary winding. Each corner of tertiary has Bushing CT. One corner say Y phase is earthed. In case of earth fault in tertiary, return path is available as one corner is earthed. Y phase CT is wired to relay for tripping.

If Bushing CT is not available, any ring type CT on the strip or cable connecting corner of delta to earth mat can be used.

In case of 3 No single phase autotransformer banks (e.g. 765/400/33kV, 3x333MVA), delta formation is done by external cables. Corner earthing is also done by single or multiple runs of 1C cables.

Sample calculation for cable sizing is given in Appendix 5-7.

5.15.4 Future Trends in Provision of Tertiary

The purpose of this winding is the compensation or suppression of 3rd harmonics caused by the magnetization current (no load current) of the transformer. But in modern transformers, with vastly superior core material, no load current has been brought down to less than 0.3%. Third harmonic component is about 30% of no load current which itself is very small. Hence the relevance of tertiary has diminished considerably. The present trend is to procure ICTs (applicable also for star-star and star- Zig Zag) without tertiary irrespective of rating of transformers. 3 limbed core type transformer is preferred. Elimination of tertiary improves reliability of transformer and reduction in cost without adverse effect on performance.

Since tertiary is eliminated, low voltage winding is not available to conduct no load test on ICT at vendor's works. It is recommended to procure the ICT without tertiary only from reputed and experienced vendors who also manufacture EHV shunt reactors. They will have EHV testing transformer set up for conducting no load test.

5.16 CT SECONDARY GROUNDING

CTs located in switchyard (AIS) are connected to CRP (Control & Relay Panel) located in control room by 2.5 mm² cu cable. The neutral formed in BMB (Bay Marshalling Box) is extended to CRP. Refer Figure 5.40.

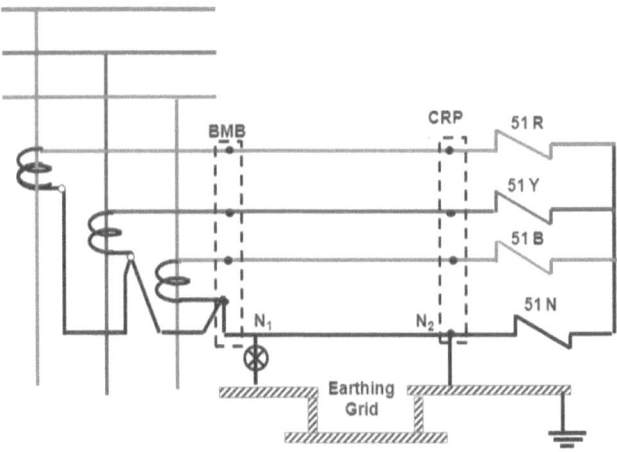

Figure 5.40

Following discussions are based on IEEE Guide on this topic [36].

The secondary of CT which is commonly accessed by testing engineer shall be safe to work. The primary of CT is at very high voltage. The stray capacitances present are:

(i) Between Primary and Secondary of CT (C_1)
(ii) Between Secondary of CT and ground (C_2)

Refer Figure 5.41.

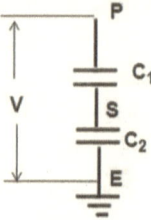

Figure 5.41

Without grounding, the circuit acts like a voltage divider. The secondary voltage to ground is given by:

$$V_{SE} = V \cdot \frac{1}{1 + \frac{C_2}{C_1}}$$

Depending on values of C_1 and C_2, the secondary voltage V_{SE} could be substantial. This results in safety hazard to testing personnel. Also the connected secondary equipment may also be damaged due to high voltage. Hence it is imperative to ground the secondary of CT to maintain it at ground potential as shown in Figure 5.42. In this case, $C_2 = \infty$ and $V_{SE} = 0$.

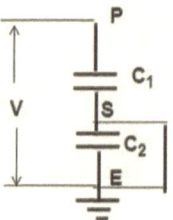

Figure 5.42

The next question is where to ground the CT secondary circuit. The important point in this context is to avoid multiple grounds. For example, in Figure 5.40, assume the grounding is done at both Bay Marshalling Box (N_1) and Control & Relay panel (N_2). Now the entire CT secondary circuit becomes part of earthing grid. In case of ground faults, part of the fault current (in kA) flowing in main earthing grid will find a point through CT secondary circuit leading to mal-operation of relays. Also CT secondary wiring may be damaged if it carries large current as its size is only 2.5 mm² cu. In conclusion, do not earth at both N_1 and N_2.

If CT secondary is to be grounded, it is preferable to ground at CRP end. This is where most of the testing is being carried out. Also if CT secondary wires are to be tested for IR (Insulation Resistance), it is necessary to isolate only one earth connection at N_2 before doing the test.

In case of GIS, distance between CRP and switchgear is not that much as in AIS. In this case grounding is done either at N_1 or N_2 as per utility practice.

Grounding Transformer Specification without Ambiguity

Appendix **5-1**

5-1.0 INTRODUCTION

It is a well-known practice to use Zig Zag transformer to ground a bus fed by an ungrounded system. Though the evaluation of Zig Zag impedance is straight forward, many times confusion arises when specifying the same to the vendor. The vendor must clearly understand what the user 'really' wants; otherwise he may supply equipment that may not meet user's requirements. Some of the finer points involved in Zig Zag transformer design, specification and testing to ensure clear understanding between the user and the vendor are clarified below [1].

5-1.1 CASE STUDY 1 (WITHOUT NGR)

Consider an ungrounded system, which is to be grounded through grounding transformer. The system voltage is 34.5kV and three phase fault level is 525 MVA. The ground fault current is to be limited to 5500A. Refer Cl A.1 of [26]

Choose Base MVA = 100 MVA
Base kV = 34.5kV
$I_{3PH} = 525/(\sqrt{3} \times 34.5) = 8.8kA$
$I_{1PH} = 5.5kA$
$K_F = 5.5/8.8 = 0.63$
Since $K_F > 0.6$, the system is effectively grounded (Refer Sec 5.4)

$Z_{base} = \dfrac{34.5^2}{100}$
= 11.9025

$I_{base} = \dfrac{100}{(\sqrt{3} \times 34.5)}$
= 1.6735 kA

I_F = 5500 A
$I_F = \dfrac{5.5}{1.6735}$
= 3.2866 pu.

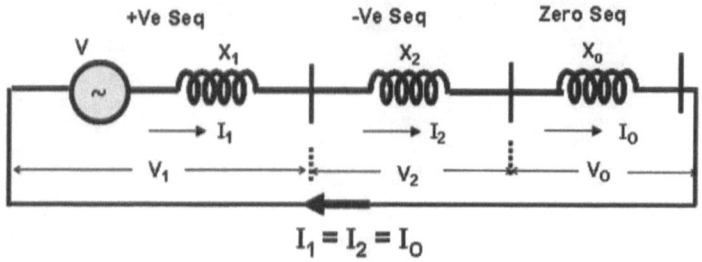

Figure 5-1.1

From theory of symmetrical components (Figure 5-1.1),

$I_F = 3I_0$

$I_0 = I_F/3$

   $= 3.2866/3$

   $= 1.0955\,pu$

I_0 in A $= I_0$ in pu x I_{Base}

   $= 1.0955 \times 1.6735$

   $= 1.833\,kA$

Positive sequence impedance of system in pu,

$X_1 = \dfrac{\text{Base MVA}}{\text{Fault MVA}}$

   $= \dfrac{100}{525}$

   $= 0.1905\,pu$

Negative sequence impedance of system, $X_2 = 0.1905\,pu$
From Figure 5-1.1,

$I_0 = \dfrac{1.0}{(X_1 + X_2 + X_0)}$

   $= \dfrac{1.0}{(0.381 + X_0)}$

   $= 1.0955\,pu$

Zero sequence impedance, $X_0 = 0.5318\,pu$
Zero sequence impedance in ohms, $X_0 = X_0$ in pu x Z_{Base}

$= 0.5318 \times 11.9025$

$= 6.3299\ \Omega\,/\,\text{phase}$ (5-1.1)

Notional 3φ rating of grounding transformer
Rating = $\sqrt{3}$ V I_0
= $\sqrt{3}$ x 34.5 x 1.833
≈ 110 MVA

On 110 MVA Base, Zero sequence impedance of grounding transformer,

$X_0 = \dfrac{110}{100}$ x 0.5318
= 0.585 pu
X_0 = 58.5 %
$Z_{base} = \dfrac{34.5^2}{110}$
= 10.8205

Zero sequence impedance, X_0 = 0.585pu
Zero sequence impedance in ohms, X_0 = X_0 in pu x Z_{Base}

$$= 0.585 \times 10.8205$$
$$= 6.3299 \ \Omega/\text{phase} \quad (5\text{-}1.2)$$

It is same as obtained previously in (5-1.1))

The cause for ambiguity and confusion arises from calculation of fault current from ohmic or percentage value (Figure 5-1.2).

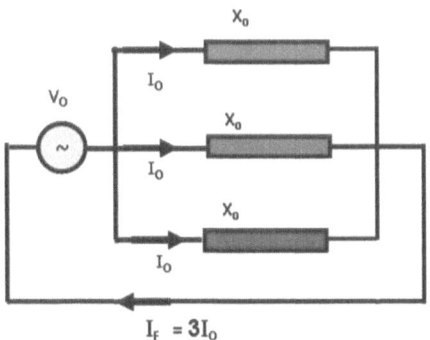

Figure 5-1.2

The common (wrong) method is as follows:

V_0 = 1.0 pu
= $\dfrac{34.5}{\sqrt{3}}$ kV
= 19.9186 kV

$$I_F = \frac{V_0}{\left(\frac{6.3299}{3}\right)} kA$$

= 9.4404 kA - which is wrong.

The mistake above is to consider the voltage, $V_0 = (34.5/\sqrt{3})$ kV

This is true only if source is infinite bus (source impedance is zero, i.e. $X_1 = X_2 \cong 0$).

From Figure 5-1.1,

$V_0 = 1.0 - \{(X_1 + X_2) I_0\}$
   = 1.0 - (0.381 x 1.0955)
   = 0.5826 pu.
$V_0 = 0.5826 \times (34.5/\sqrt{3})$
   = 11.6046 kV

The correct method for current calculation (Figure 5-1.2):

$$I_F = \frac{V_0}{\left(\frac{6.3299}{3}\right)} kA$$

$$= \frac{11.6046}{\left(\frac{6.3299}{3}\right)}$$

= 5.5 kA - which is correct

If the user specifies only voltage (34.5kV) and the fault current as (5.5kA), the vendor will assume infinite bus and will offer

$X_0 = (34.5/\sqrt{3})/(5.5/3)$
   = 10.8647 Ω/phase

The above value is wrong.

If the Zig Zag transformer with above value is connected to the actual system, the fault current will be much less than design value of 5.5kA as can be seen here:

$X_0 = \frac{10.8647}{11.9025}$ on 100 MVA Base
   = 0.9128 pu

$I_0 = \frac{1.0}{(X_1 + X_2 + X_0)}$

$= \frac{1.0}{(0.381 + 0.9128)}$

= 0.7729 pu

$I_F = 3I_0$
   = 3 x 0.7729
   = 2.3187 pu

I_0 in A = I_0 in pu x I_{Base}
  = 2,3187 x 1.6735
  = 3.88kA

This is less than desired value of 5.5 KA

Ground fault relay co-ordination may be affected in this case.

Hence, to avoid confusion and add clarity to the specification, the following may be included in the specification:

Type: Zig Zag grounding transformer

Voltage: 34.5kV

Rating: 110 MVA for 10 sec

Zero sequence reactance X_0 = 58.5% (6.3299 Ω/Phase)

We saw earlier that when voltage V_0 = 11.6046kV is applied, the resultant neutral current is 5.5 kA.

During shop testing, (Figure 5-1.2) when Zero sequence (single phase) voltage of 230V is applied, the neutral current shall be:

$$I_N = \frac{5500 \times 0.230}{11.6046}$$
  = 109 A

5-1.2 CASE STUDY 2 (WITH NGR)

The system shown for simulation is shown in Figure 5-1.3.

The 415V generator is connected to 11kV system through star – delta transformer. It is now proposed to ground 11kV system through Zig Zag grounding transformer to limit ground fault current to within 100A.

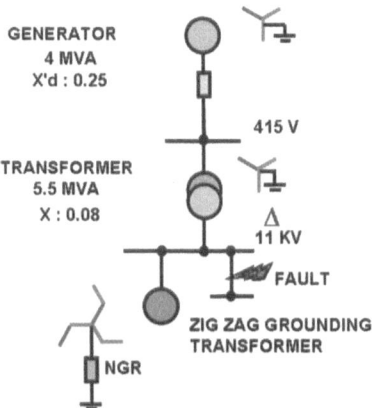

Figure 5-1.3

Choose 10 MVA base.

$Z_{Base} = 11^2/10 = 12.1 \Omega$
$I_{Base} = 10/(\sqrt{3} \times 11) = 524.8 A$
$I_F = 100 A$
I_F in pu $= 100/524.8$
$\qquad = 0.1906$ pu
$I_0 = I_F/3$
$\quad = 0.1906/3$
$\quad = 0.0635$ pu

Converting generator and transformer impedances on 10 MVA base

$Z_G = 0.25 \times \left(\dfrac{10}{4}\right)$
$\quad = 0.625$ pu

$Z_T = 0.08 \times \left(\dfrac{10}{5.5}\right)$
$\quad = 0.1455$ pu

$Z_1 = Z_2$
$\quad = 0.625 + 0.1455$
$\quad = 0.7705$ pu

From Figure 5-1.1,

$I_0 = \dfrac{1.0}{(Z_1 + Z_2 + Z_0)}$
$\quad = \dfrac{1.0}{(1.541 + Z_0)}$
$\quad = 0.0635$ pu

$Z_0 = 14.2070$ pu

It is not possible to use only grounding transformer to obtain Z_0 of 1420%. The other feasible alternative is to connect a resistor in the neutral. Since current is limited to only 100A resistor is suggested. It is obvious that compared to requirement of Z_0, the values of Z_1 and Z_2 are negligible and can be ignored.

From symmetrical component theory, if resistor R is connected in the neutral circuit, $Z_0 = 3 R_N$.

$$R_N = \frac{Z_0}{3}$$

$$R_N = \frac{14.2070}{3}$$

$$= 4.7357 \text{ pu}$$

R_N in Ω = R_N in pu $\times Z_{base}$

$\qquad = 4.7357 \times 12.1 \, \Omega$

$\qquad = 57.3 \, \Omega$

In this case, as $Z_0 >> Z_1$ or Z_2, entire voltage is dropped only in zero sequence network (Figure 5-1.1) compared to only 58.26% in Case Study 1.

Summarising,

Notional rating of Zig Zag transformer = $\sqrt{3} \times 11 \times \left(\frac{100}{3}\right)$

$\qquad\qquad\qquad\qquad\qquad\qquad\qquad = 635\text{kVA for 10 sec.}$

It is customary to choose X_0 reactance value as typically $3X_1$

$X_0 = 3 \times 0.7705 = 2.3115$ pu on 10 MVA base.

On 635kVA base, reactance

$$X_0 = 2.3115 \times \left(\frac{0.635}{10}\right)$$

$\qquad = 0.1468 \text{ pu}$

$\qquad \cong 15\%$

$X_0 = 2.3115 \times 12.1 = 27.9692 \Omega/\text{phase}$

During testing, when Zero sequence (single-phase) voltage of 230V is applied, the neutral current shall be (Figure 5-1.2),

$$I_N = \frac{230}{\left(27.9692/3\right)}$$

$\qquad = 24.7 \text{ A}$

The Neutral grounding resistor R_G= 57.3 Ω and rated for $11/\sqrt{3}$kV.

Ignoring reactance of NGT, approximate value of fault current (Refer Sec 5.11.2.1) can be found as

$I_F = (11,000/\sqrt{3})/57.3 = 111$A which is nearly equal to desired value of 100A.

NGT with/without NGR

Appendix 5-2

5-2.0 INTRODUCTION

Zig Zag Neutral Grounding transformer is used to ground ungrounded system. Normally confusion does not arise when the ground fault current has to be limited to the following:
  (i) High value (say more than 60% of three phase fault current for effectively grounded system). In this case, only NGT with appropriate zero sequence reactance will suffice. Refer Sec 5-1.1.
  (ii) Low value (say less than 300A). In this case, NGT with Neutral Grounding Resistor will be the practical choice. Refer Sec 5-1.2.

However when the ground fault current has to be limited to an intermediate value like 2 to 3 kA, provision of NGR shall be critically examined. This is illustrated with an example [11].

5-2.1 SYSTEM DATA

The system data:

Transformer rating: HV/IV/LV: 400kV/132kV/33kV, YNynΔ

The objective is to provide grounding for system connected to delta winding so that earth fault current on 33kV side is limited to 2000 A.

The data is given below.

Impedance on 250 MVA base:
HV – IV: 0.10 pu
HV – LV: 0.17 pu
IV – LV: 0.08 pu
400kV fault level = 50 kA

Analysis:
Base MVA = 250
Base Voltage = 33kV

Base Impedance, $Z_{BASE} = \dfrac{33^2}{250}$

$= 4.356 \, \Omega$

Base Current $= 250/(\sqrt{3} \times 33)$

$= 4374A$

400kV fault level $= \sqrt{3} \times 400 \times 50$

$= 34{,}640 \text{ MVA}$

$X_{SYS} = \dfrac{250}{34{,}640}$

$= 0.0072 \text{pu}$

$X_1 = 0.17 + 0.0072$

$= 0.1772 \text{pu}$

$X_2 = X_1$

5-2.2 CASE 1 - NGT WITHOUT NGR

The equivalent circuit is shown in Figure 5-2.1. NGR is ignored in this case.

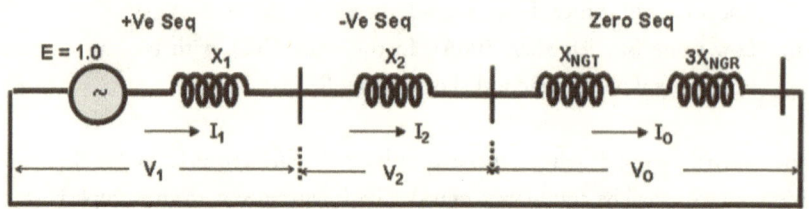

Figure 5-2.1

$I_F = 2000A$

$= 2000/4374$

$= 0.4573 \text{pu}$

$I_0 = I_F/3$

$= 0.1524 pu$

$I_0 = \dfrac{1.0}{\left(X_1 + X_2 + X_0^T \right)}$

$X_1 + X_2 + X_0^T = 6.5617$

$X_0^T = 6.5617 - 0.1772 - 0.1772$

$= 6.2073 \text{pu}$

$= 6.2073 \times 4.356$

$= 27.04 \, \Omega$

Notional Rating of NGT =
$$= \sqrt{3} \, V \, I_0$$
$$= \sqrt{3} \times 33 \times \left(\frac{2000}{3}\right)$$
$$= 38.1 \text{ MVA}$$
$$= \text{say } 40 \text{ MVA}$$

Base Impedance, $Z_{Base} = 33^2/40$
$$= 27.23 \, \Omega$$

$$X_0^T = \left(\frac{27.04}{27.23}\right) \times 100$$
$$= 100\%$$

For grounding transformer, 100% impedance is not an abnormal figure and is achievable.

5-2.3 CASE 2 - NGT WITH NGR
The impedances are worked out considering NGR.

Desired zero sequence impedance to limit ground fault current within 2000A is
$X_0^T = 6.2073$ pu

It is customary to assume Zero sequence reactance of NGT to be $3X_1$.

$X_0^{NGT} = 3 \times 0.1772$
$= 0.5316$ pu
$= 0.5316 \times 4.356$
$= 2.3157 \Omega$

Zero sequence reactance of reactor = X_0^R

$X_0^T = X_0^{NGT} + 3X_0^R$ (Refer Figure 5-2.1)

$$X_0^R = \frac{(6.2073 - 0.5316)}{3}$$
$= 1.8919$ pu
$= 1.8919 \times 4.356$
$= 8.2411 \Omega$

5-2.4 REASON FOR CHOOSING ONLY NGT AND NOT NGT & NGR
The grounding transformer reactance is 27Ω without NGR and 2.3Ω with NGR. It is not economical to design a transformer with very low reactance value as in the case with NGR for following reasons.

The reactance is directionally proportional to T^2. If the reactance value is low, the number of turns (T) will be low. For a given applied voltage, volts per turn (V/T) will be high and hence flux (φ) will be high.

$$\frac{V}{T} = \varphi = 4.44 \times f \times B \times A$$

For a given flux density B (say, 1.6 Tesla), area of cross section (A) will be high. The reactance is inversely proportional to coil height H. This has to be increased to get lower reactance. The core frame height also increases correspondingly. Because of the above two reasons, the core weight and core loss are substantially higher for grounding transformer with lower reactance. The differential cost in capex would be about 20 to 25% for transformers between low reactance (2.3Ω/phase) and high reactance (27Ω/phase). Moreover, NGR cost will be additional.

Also the core loss with high reactance transformer will be lower by 4KW compared to low reactance transformer. The major loss in grounding transformer is only core loss (as it does not carry load current under normal conditions), The opex in a life time of 30 years will be substantial.

In conclusion, whenever the ground fault current is to be limited to say 2 to 3 kA, which is neither too low nor too high, it is prudent to have preliminary discussions with vendors before finalising NGR along with NGT.

Sizing of NGT

Appendix **5-3**

5-3.0 ANALYSIS

Zig Zag Neutral Grounding transformer is used to ground to ungrounded system.
Refer Figure 5-3.1

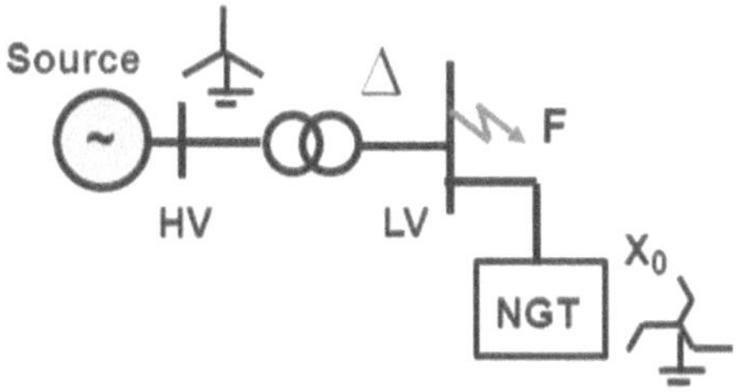

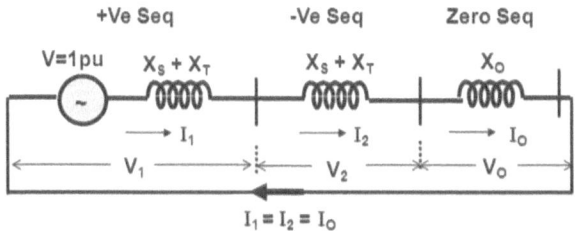

Figure 5-3.1

System Impedance	$= X_S$
Transformer Impedance	$= X_1 = X_2 = X_T$
NGT Impedance	$= X_0$

Sequence network for (L-G) on LV side is shown in Figure 5-3.2

$$I_1 = I_2 = I_0$$

Figure 5-3.2

Positive Sequence Impedance
$$X_1 = X_S + X_T$$
Negative Sequence Impedance
$$X_2 = X_1$$
Three phase Fault Current
$$I_{3P} = \frac{1}{X_1}$$

Desired Single phase fault current ($0.6 < K_F < 1.0$)
$$I_{1P} = K_F I_{3P} = 3I_0$$

$$I_0 = \frac{I_{1P}}{3}$$

$$I_0 = I_1 = I_2 = \frac{1.0}{X_1 + X_2 + X_0}$$

$$= \frac{1.0}{2X_1 + X_0}$$

$$X_0 = \frac{1}{I_0} - 2X_1 \tag{5-3.1}$$

5-3.1 OPEN DELTA VOLTAGE

$$V_1 = 1.0 - I_0 X_1 \tag{5-3.2}$$
$$V_2 = -I_2 X_2$$
$$= -I_0 X_1 \tag{5-3.3}$$
$$V_0 = -I_0 X_0$$

From Equation (5-3.2) and (5-3.3)
$$V_1 - V_2 = 1.0 \tag{5-3.4}$$
$$V_a = V_0 + V_1 + V_2$$
$$= 1.0 - I_0(2X_1 + X_0) \tag{5-3.5}$$

From Equation (5-3.1) and (5-3.5)
$$V_a = 0$$
V_a is the faulted phase
$$V_b = V_0 + a^2 V_1 + a V_2$$
$$= V_0 + \left(-0.5 - j\frac{\sqrt{3}}{2}\right)V_1 + \left(-0.5 + j\frac{\sqrt{3}}{2}\right)V_2$$
$$= [V_0 - 0.5(V_1 + V_2)] - j\frac{\sqrt{3}}{2}(V_1 - V_2)$$

From Equation (5-3.4)

$$V_b = [V_0 - 0.5(V_1 + V_2)] - j\frac{\sqrt{3}}{2}$$
$$= |V| \angle -\theta$$
$$V_c = V_0 + aV_1 + a^2V_2$$
$$= V_0 + \left(-0.5 + j\frac{\sqrt{3}}{2}\right)V_1 + \left(-0.5 - j\frac{\sqrt{3}}{2}\right)V_2$$
$$= [V_0 - 0.5(V_1 + V_2)] + j\frac{\sqrt{3}}{2}(V_1 - V_2)$$
$$= [V_0 - 0.5(V_1 + V_2)] + j\frac{\sqrt{3}}{2}$$
$$= |V| \angle \theta$$

Open Delta Voltge
$$V_D = V_a + V_b + V_c$$
$$= 2|V| \cos\theta$$

Coefficient of Earthing (COE) is similar to E_F except that over voltage on healthy phase is scaled with respect to rated line voltage instead of phase voltage. Hence, for effectively grounded system,

$$COE \leq (1.4/\sqrt{3})$$
$$\leq 0.8$$

5-3.2 SPREAD SHEET CALCULATION
Refer Table 5-3.1

Input: B2 to B9
Output: B44 to B48
Testing of NGT is done at low voltage (230V to 440V) at works. The neutral current is calculated in two ways.

(i) Test voltage applied = V_T

$$\text{Neutral Currnt} = \left(\frac{I_{1P}}{V_0}\right)V_T$$

(ii) Using circuit analysis (Figure 5-3.3)

$$\text{Neutral Currnt} = \left(\frac{V_T}{\left(\frac{X_0}{3}\right)}\right) = \frac{3V_T}{X_0}$$

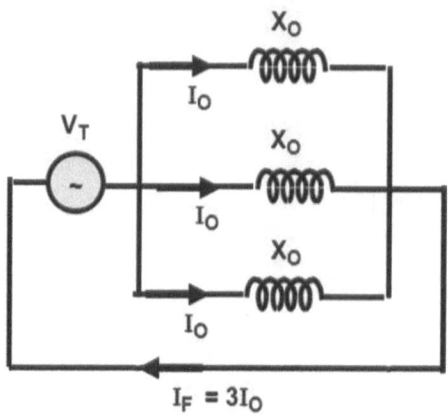

Figure 5-3.3

5-3.3 EXAMPLE

220kV Source Fault Level = 7500 MVA
Transformer = 220/33kV, 100 MVA, 15%, Star – Delta
33kV bus is grounded through NGT.
Find the reactance of NGT to limit ground fault current to 70% of 3 phase fault current. Find resulting COE (Coefficient of Earthing) and open delta voltage.

The solution is given in Table 5-3.2.
  Factor K_F (B9) can be varied from 0.01 to 1.0.
  The resulting voltage on healthy phases (B36) and open delta voltage (B40) can be obtained. Curves similar to those given in Figure 5.32 can be obtained.

Table 5-3.1: NGT Sizing

No.	A	B	C
1	Evaluation of Zig Zag Grounding Transformer Parameters		
2	HV side Rated Voltage of transformer	220	kV
3	LV side Rated Voltage of transformer	33	kV
4	Transformer Rating	100	MVA
5	Transformer Positive Sequence Reactance = XP	15	%
6	HV winding connection	Star	Fixed
7	LV winding connection	Delta	Fixed
8	HV side system Fault Level	7500	MVA
9	Min 1Phase/3 Phase fault current factor KF	0.7	
10	Base MVA	=B4	MVA
11	Base kV=	=B3	kV
12	Base Impedance = Base_kV^2/Base MVA	=B11*B11/B10	Ohms
13	Base Current = Base MVA/(1.732* Base kV)	=B10/(1.732*B11)	kA
14	Source Impedance in pu = Base MVA/Fault MVA = XS	=B10/B8	pu
15	Transformer XP = XN	=B5/100	pu
16	Three phase fault level on HV side of transformer	=B8/(1.732*B2)	kA
17	Three phase fault level on LV side of transformer	=B10/(B14+B15)	MVA
18	Three phase fault level on LV side of transformer	=B17/(1.732*B11)	kA
19	Desired single phase fault level on LV side of transformer	=B18*B9	kA
20	Desired single phase fault level on LV side of transformer	=B19/B13	pu
21	IO = IF/3 = 1/(2(XS+XP)+XO_NGT)	=B20/3	pu
22	2(XS+XP)+ X0_NGT	=1/B21	pu
23	XO_NGT	=B22-(2*(B14+B15))	pu
24	XO_NGT	=B23*B12	Ω/Phase
25	XPOS=XS+XP	=B14+B15	pu
26	XNEG=XS+XP	=B14+B15	pu
27	XZER	=B23	pu

Table 5-3.1: NGT Sizing (Contd)

No.	A	B	C
28	I1=IO	=B21	pu
29	I2=IO	=B21	pu
30	V1 = 1.0 – (I1 x XPOS)	=1 – (B28*B25)	pu
31	V2=-I2*XNEG	=-B29*B26	pu
32	VO = -IO*XZER	=-B21*B27	pu
33	VA = V1+V2+Vo Faulted Phase	=B30+B31+B32	pu
34	VB(Real) = -0.5(V1+V2)+ VO	=(-0.5*(B30+B31))+B32	pu
35	VB(Imag)	=-SQRT(3)/2	pu
36	VB(Mag) Unfaulted Phase	=SQRT(B34*B34+B35*B35)	pu
37	VB(Ang)	=ATAN2(B34,B35)*180/3.1417	deg
38	VC(Mag) Unfaulted Phase	=B36	pu
39	VC(Ang)	=-B37	deg
40	Open Delta Voltage:	=2*B36*COS(RADIANS(180-B39))	pu
41	Open Delta Voltage:	=B40*B3/1.732	kV
42	Coefficient of Earthing (COE)	=B36/1.732	COE<0.8
43	**Recommended rating of NGT**		
44	Rated Voltage	=B3	kV
45	Type of connection	Zig Zag	Fixed
46	Rated Fault current (Neutral Current)	=B19	kA
47	Zero sequence reactance per phase	=B24	Ohms
48	Fault current duration	10 Sec	Fixed
49	Shop testing of NGT with 1Phase LT supply		
50	Test Voltage between shorted terminals & Neutral	300	Volts
51	VO	=ABS(B32)*B3/1.732	kV
52	Resulting Neutral current	=(B19/B51)*B50	Amps
53	Resulting Neutral current Verification	=3*B50/B24	Amps

Table 5-3.2: NGT Sizing - Example

No.	A	B	C
1	Evaluation of Zig Zag Grounding Transformer Parameters		
2	HV side Rated Voltage of transformer	220	kV
3	LV side Rated Voltage of transformer	33	kV
4	Transformer Rating	100	MVA
5	Transformer Positive Sequence Reactance = XP	15	%
6	HV winding connection	Star	Fixed
7	LV winding connection	Delta	Fixed
8	HV side system Fault Level	7500	MVA
9	Min 1Phase/3 Phase fault current factor KF	0.7	
10	Base MVA	100	MVA
11	Base kV=	33	kV
12	Base Impedance = Base_kV^2/Base MVA	10.8900	Ohms
13	Base Current = Base MVA/(1.732* Base kV)	1.7496	kA
14	Source Impedance in pu = Base MVA/Fault MVA = XS	0.0133	pu
15	Transformer XP = XN	0.15	pu
16	Three phase fault level on HV side of transformer	19.68	kA
17	Three phase fault level on LV side of transformer	612	MVA
18	Three phase fault level on LV side of transformer	10.7118	kA
19	Desired single phase fault level on LV side of transformer	7.4983	kA
20	Desired single phase fault level on LV side of transformer	4.2857	pu
21	IO = IF/3 = 1/(2(XS+XP)+XO_NGT)	1.4286	pu
22	2(XS+XP)+ X0_NGT)	0.7000	pu
23	XO_NGT	0.3733	pu
24	XO_NGT	4.0656	Ω/Phase
25	XPOS=XS+XP	0.1633	pu
26	XNEG=XS+XP	0.1633	pu
27	XZER	0.3733	pu

Table 5-3.2: NGT Sizing - Example (Contd)

No.	A	B	C
28	I1=IO	1.4286	pu
29	I2=IO	1.4286	pu
30	V1 = 1.0 − (I1 x XPOS)	0.7667	pu
31	V2=-I2*XNEG	-0.2333	pu
32	VO = -IO*XZER	-0.5333	pu
33	VA = V1+V2+Vo Faulted Phase	0.0000	pu
34	VB(Real) = -0.5(V1+V2)+ VO	-0.8000	pu
35	VB(Imag)	-0.8660	pu
36	VB(Mag) Unfaulted Phase	1.1790	pu
37	VB(Ang)	-132.7	deg
38	VC(Mag) Unfaulted Phase	1.1790	pu
39	VC(Ang)	132.7	deg
40	Open Delta Voltage:	1.60	pu
41	Open Delta Voltage:	30.48	kV
42	Coefficient of Earthing (COE)	0.6807	COE<0.8
43	**Recommended rating of NGT**		
44	Rated Voltage	33	kV
45	Type of connection	Zig Zag	Fixed
46	Rated Fault current (Neutral Current)	7.4983	kA
47	Zero sequence reactance per phase	4.0656	Ohms
48	Fault current duration	10 Sec	Fixed
49	Shop testing of NGT with 1Phase LT supply		
50	Test Voltage between shorted terminals & Neutral	300	Volts
51	VO	10	kV
52	Resulting Neutral current	221	Amps
53	Resulting Neutral current Verification	221	Amps

Sizing of NGR

Appendix 5-4

5-4.0 INTRODUCTION

Neutral Grounding Reactor (NGR) is introduced between Neutral and Ground to limit the ground fault current to desired value. The popular application is in sub-transmission network for $Y_N z_n$ or Dz_n transformer. Refer Figure 5-4.1 used in simulation studies.

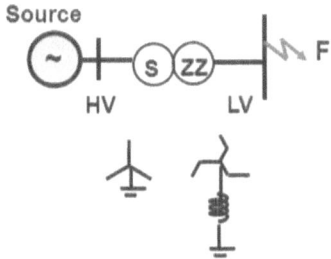

Figure 5-4.1

Procedure for sizing NGR connected between Zig Zag Neutral and ground is very similar to that of NGT given in Appendix 5-3. However there are two minor differences:

(i) In case of NGT sizing (Appendix 5-3), zero sequence impedance of power transformer (X_0) is not involved in fault level calculation. In case of NGR sizing, X_0 of power transformer is part of calculation procedure. In case of Y-Δ or Y-Y $X_1 = X_2 \cong X_0$. (Say 10% to 15%).
In case of Y- Zig Zag, X_0 is much smaller and typically about 2% measured from Zig Zag side.

(ii) With NGR, Zero sequence impedance up to the point of fault is given by.
$X_0 = X_0^T + 3X_{NGR}$
X_0^T: Zero sequence impedance of power transformer.
X_{NGR}: Reactance of NGR.

5-4.1 SPREAD SHEET CALCULATION

Refer Table 5-4.1

Input: B2 to B10

Output: B44 to B50

It can be used for following vector groups: $Y_N z_n$, Dz_n, Dy_n

5-4.2 EXAMPLE:

Source fault level: 7500 MVA

Transformer 220/33kV, $Y_N z_n$, 125 MVA, $Z_1 = Z_2 = 15\%$; $Z_0 = 2\%$;
Zig Zag Neutral is grounded through NGR.

Determine the value of NGR to limit ground fault current to 60% of 3 phase fault current.

Determine associated COE and open delta Voltage.

The solution is given in Table 5-4.2

Table 5-4.1: Neutral Grounding Reactor Sizing Calculation

No.	A	B	C
1	Calculation for NGR value for Star - Zig Zag or Delta – Zig Zag Transformer		
2	HV side Rated Voltage of transformer	220	kV
3	LV side Rated Voltage of transformer	33	kV
4	Transformer Rating	125	MVA
5	HV winding connection – Star or Delta	Star	
6	LV winding connection	Zig Zag	
7	Transformer Positive Sequence Reactance = XP	15	%
8	Transformer Zero Sequence Reactance = XO (<<XP)	2	%
9	HV side system Fault Level	7500	MVA
10	Min 1Phase/3 Phase fault current factor KF	0.6	0.6<Factor<1
11	Base MVA	=B4	MVA
12	Base kV	=B3	kV
13	Base Impedance = Base_kV^2/Base MVA	=B12*B12/B11	Ohms
14	Base Current = Base MVA/(1.732* Base kV)	=B11/(1.732*B12)	kA
15	Source Impedance in pu = Base MVA/Fault MVA = XS	=B11/B9	pu
16	Transformer XP = XN	=B7/100	pu
17	Transformer XO	=B8/100	pu
18	Three phase fault level on HV side of transformer	=B9/(1.732*B2)	kA
19	Three phase fault level on LV side of transformer	=B11/(B15+B16)	MVA
20	Three phase fault level on LV side of transformer	=B19/(1.732*B12)	kA
21	Desired single phase fault level on LV side of transformer	=B20*B10	kA
22	Desired single phase fault level on LV side of transformer	=B21/B14	pu
23	IO = IF/3 = 1/(2(XS+XP)+XOO)	=B22/3	pu
24	2(XS+XP)+XOO	=1/B23	pu
25	XOO = XO + 3 XNGR	=B24-(2*(B15+B16))	pu
26	XNGR	=(B25-B17)/3	pu
27	XNGR – Reactor connected to Secondary Neutral	=B26*B13	Ω

Table 5-4.1: Neutral Grounding Reactor Sizing Calculation (Contd)

No.	A	B	C
28	XPOS=XS+XP	=B15+B16	pu
29	XNEG=XS+XP	=B15+B16	pu
30	XZER=XO +3*XNGR	=(B8/100)+3*B26	pu
31	I1=IO	=B23	pu
32	I2=IO	=B23	pu
33	V1 = 1.0 – (I1 x XPOS)	=1 – (B31*B28)	pu
34	V2=-I2*XNEG	=-B32*B29	pu
35	VO = -IO*XZER	=-B23*B30	pu
36	VA = V1+V2+Vo Faulted Phase	=B33+B34+B35	pu
37	VB(Real) = -0.5(V1+V2)+ VO	=(-0.5*(B33+B34))+B35	pu
38	VB(Imag)	=-SQRT(3)/2	pu
39	VB(Mag) Unfaulted Phase	=SQRT (B37*B37+B38*B38)	pu
40	VB(Ang)	=ATAN2 (B37,B38)*180/3.1417	deg
41	VC(Mag) Unfaulted Phase	=B39	pu
42	VC(Ang)	=-B40	Deg
43	Open Delta Voltage:	=2*B39*COS (RADIANS(180-B42))	pu
44	Open Delta Voltage:	=B43*B3/1.732	kV
45	Coefficient of Earthing (COE)	=B41/1.732	COE<0.8
46	**Recommended rating of NGR**		
47	Rated Voltage	=B3	kV
48	Rated Current	=B21	kA
49	Fault Current Duration – Fixed	10	Sec
50	Impedance	=B27	Ω

Sizing of NGR • **183**

Table 5-4.2: Neutral Grounding Reactor Sizing Calculation – Example

No.	A	B	C
1	**Calculation of NGR value for Star – Zig Zag or Delta – Zig Zag Transformer**		
2	HV side Rated Voltage of transformer	220	kV
3	LV side Rated Voltage of transformer	33	kV
4	Transformer Rating	125	MVA
5	HV winding connection – Star or Delta	Star	
6	HV winding connection	Zig Zag	
7	Transformer Positive Sequence Reactance = XP	15	%
8	Transformer Zero Sequence Reactance = X0 (<<XP)	2	%
9	HV side system Fault Level	7500	MVA
10	Min 1Phase/3 Phase fault current factor KF	0.6	0.6<Factor<1
11	Base MVA	125	MVA
12	Base kV	33	kV
13	Base Impedance = Base_kV^2/Base MVA	8.7120	Ohms
14	Base Current = Base MVA/(1.732* Base kV)	2.1870	kA
15	Source Impedance in pu = Base MVA/Fault MVA = XS	0.0167	pu
16	Transformer XP = XN	0.15	pu
17	Transformer XO	0.02	pu
18	Three phase fault level on HV side of transformer	19.68	kA
19	Three phase fault level on LV side of transformer	750	MVA
20	Three phase fault level on LV side of transformer	13.1220	kA
21	Desired single phase fault level on LV side of transformer	7.8732	kA
22	Desired single phase fault level on LV side of transformer	3.6000	pu
23	IO = IF/3 = 1/(2(XS+XP)+XOO)	1.2000	pu
24	2(XS+XP) + XOO	0.8333	pu
25	XOO=XO + 3 XNGR	0.5000	pu
26	XNGR	0.1600	pu
27	XNGR – Reactor connected to Secondary Neutral	1.3939	Ω

Table 5-4.2: Neutral Grounding Reactor Sizing Calculation – Example (Contd)

No.	A	B	C
28	XPOS = XS + XP	0.1667	pu
29	XNEG = XS + XP	0.1667	pu
30	XZER = XO + 3* XNGT	0.5000	pu
31	I1=IO	1.2000	pu
32	I2=IO	1.2000	pu
33	V1 = 1.0 – (I1 x XPOS)	0.8000	pu
34	V2=-I2*XNEG	-0.2000	pu
35	VO = -IO*XZER	-0.6000	pu
36	VA = V1+V2+Vo Faulted Phase	0.0000	pu
37	VB(Real) = -0.5(V1+V2) + VO	-0.9000	pu
38	VB(Imag)	-0.8660	pu
39	VB(Mag) Unfaulted Phase	1.2490	pu
40	VB(Ang)	-136.1	deg
41	VC(Mag) Unfaulted Phase	1.2490	pu
42	VC(Ang)	136.1	deg
43	Open Delta Voltage:	1.80	pu
44	Open Delta Voltage:	34.29	kV
45	Coefficient of Earthing (COE)	0.72	COE<0.8
46	**Recommended rating of NGR**		
47	Rated Voltage	33	kV
48	Rated Current	7.8732	kA
49	Fault current Duration –Fixed	10	Sec
50	Impedance	1.3939	Ω

Zig Zag – Star Transformer for Auxiliary Supply in Switchyard

Appendix 5-5

5-5.0 INTRODUCTION

EHV switchyards are located sometimes at remote places where LT auxiliary supply may not be available from nearby sources. The switchyard may have ICTs (e.g. 765/400/33kV). 33kV winding is delta. If Zig Zag grounding transformer is envisaged on 33kV bus, it is possible to add auxiliary winding to derive LT supply. The feasibility of supplying single phase loads from Zig Zag – Star transformer is analysed below [11].

5-5.1 LOAD CONNECTED BETWEEN PHASES OF STAR WINDING

Refer Figure 5-5.1.

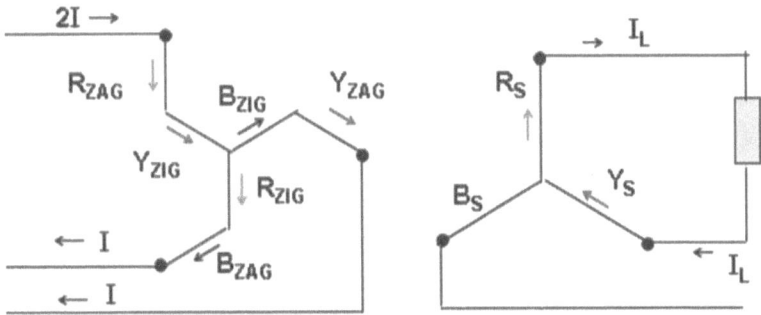

Figure 5-5.1

AT balance is obtained when the when load is connected line to line. ATs due to currents in R_{ZAG} and R_{ZIG} are balanced by load current flowing in R_S. Similarly ATs due to currents in Y_{ZIG} and Y_{ZAG} are balanced by returning load current flowing in Y_S. ATs due to current in B_{ZAG} and B_{ZIG} neutralize each other and balancing current from star side (B_S) is not required. Hence it is feasible to draw single phase load in this case. If the required single phase voltage is 240V, the star winding shall be rated for line voltage of 240V. The single phase loads will be distributed among RY, YB and BR.

5-5.2 LOAD CONNECTED BETWEEN PHASES AND NEUTRAL OF STAR WINDING

Refer Figure 5-5.2.

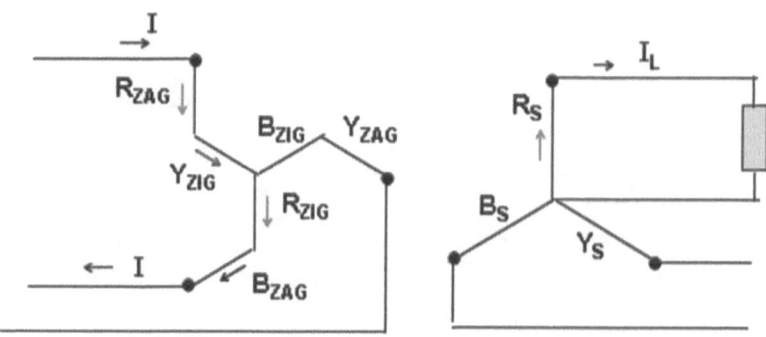

Figure 5-5.2

ATs due to currents in R_{ZAG} and R_{ZIG} can be balanced by load current flowing in R_S. Since Y and B phase star winding are open, very little current is expected to flow through B_{ZIG} and Y_{ZAG}. The reflected load current is forced to return through Y_{ZIG} and B_{ZAG}. Since the corresponding star windings are open, the current on Zig Zag side behaves like exciting current or magnetizing current. The magnetizing impedance is very high. Hence the current that can be delivered is very low and even if some current is forced to flow, the core will be saturated. Regulation also will be too poor. In summary, it is not practical to supply single phase load connected between phase and neutral of star winding. If single phase loads are connected between phase and neutral, distributed in all three phases and if phase balancing is achieved with minimum neutral current flow, regulation can be normal.

Zig Zag Grounding Transformer – Short Circuit Testing

Appendix 5-6

5-6.0 INTRODUCTION

As a routine test, this is performed to find zero sequence impedance of transformer. As a type test it is performed in testing laboratories at rated voltage to prove the withstand capacity of transformer to carry the design current for the specified duration. Short circuit test for NGT is done in two ways as per Cl 10.9.8 of IEC 60076 -6 [35].

5-6.1 ALTERNATIVE 1

The earthing transformer is connected to a single phase supply between the three line terminals connected together and the neutral terminal. Refer Figure 5-6.1.

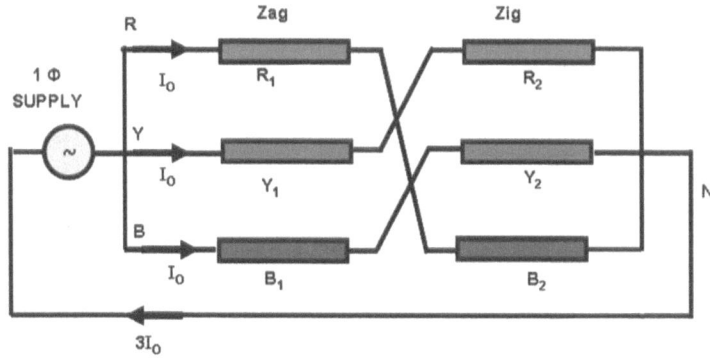

Figure 5-6.1: Alternative 1

The source in testing laboratory shall have the capacity to deliver rated ground fault current $3I_0$, where I_0 is the current circulated in each winding.

The tests done at vendor's works or at site (described in Appendix 5-1) are conceptually same except that the test voltage and current are much lower.

5-6.2 ALTERNATIVE 2

The earthing transformer is connected to a balanced three phase supply. One line terminal and neutral is shorted. Refer Figure 5-6.2.

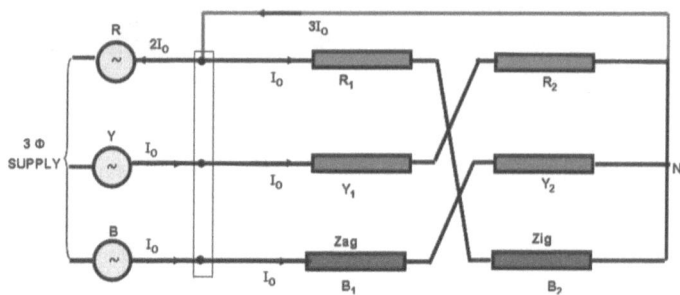

Figure 5-6.2: Alternative 2

There will be equal current (I_0) in all the windings. The current in short circuited link is $3I_0$. However the maximum current drawn from source is only $2I_0$ compared to $3I_0$ in Alternative 1. This is the reason why the testing laboratories prefer to perform short circuit testing at full voltage adopting Alternative 2 as the current source can be of lower capacity.

5-6.3 ALTERNATIVE 2 – THEORY

The theory behind this testing method is given below [11].

By inspection of Figure 5.6-2, following three voltage relationship are derived:

$B_2 - R_1 = 0$ (5-6.1)

$R_2 - Y_1 = NY = RY$ (5-6.2)

$Y_2 - B_1 = NB = RB$ (5-6.3)

Voltage equations for the Zig and Zag winding on each core are related to impedance drop as follows (Figure 5-6.3):

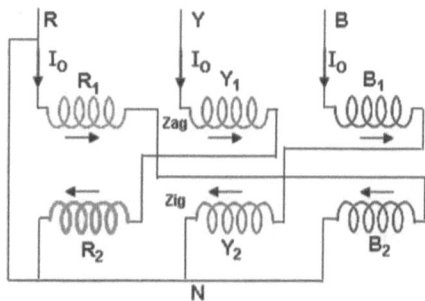

Figure 5-6.3: Winding Connection

$$R_2 - R_1 = I_0 Z_0 \qquad (5\text{-}6.4)$$
$$Y_2 - Y_1 = I_0 Z_0 \qquad (5\text{-}6.5)$$
$$B_2 - B_1 = I_0 Z_0 \qquad (5\text{-}6.6)$$

Z_0 is the leakage impedance between Zig and Zag winding.

The phasor diagram for voltages that satisfies Eqns. (5-6.1) to (5-6.6) is shown in Figure 5-6.4. The voltages present under normal conditions in the six windings are RD (R_1^o), DN (B_2^o), YE (Y_1^o), EN (R_2^o), BF (B_1^o) and FN (Y_2^o).

Assuming R – N is shorted, N collapses into R. The voltages after shorting are indicated without superscript 0. The knees of Zig Zag are displaced by equal amount (D to H, E to J and F to K).

Eqn (5-6.4) is satisfied if JR and HR (=NJ) are vectorialy subtracted to give RN which is the $I_0 Z_0$ drop which in turn is equal to phase voltage. Incidentally, the resulting current after short circuit follows from this equation, i.e. $I_0 = V_{Phase}/Z_0$.

Eqn (5-6.5) is satisfied if RK and YJ (=NK) are vectorialy subtracted to give RN which is $I_0 Z_0$ drop.

Eqn (5-6.6) is satisfied if HR (=KG) and BK are vectorialy subtracted to give BG (= RN) which is $I_0 Z_0$ drop.

The currents in lines will be in the ratio of 1:1:2. as given in Figure 5-6.2.

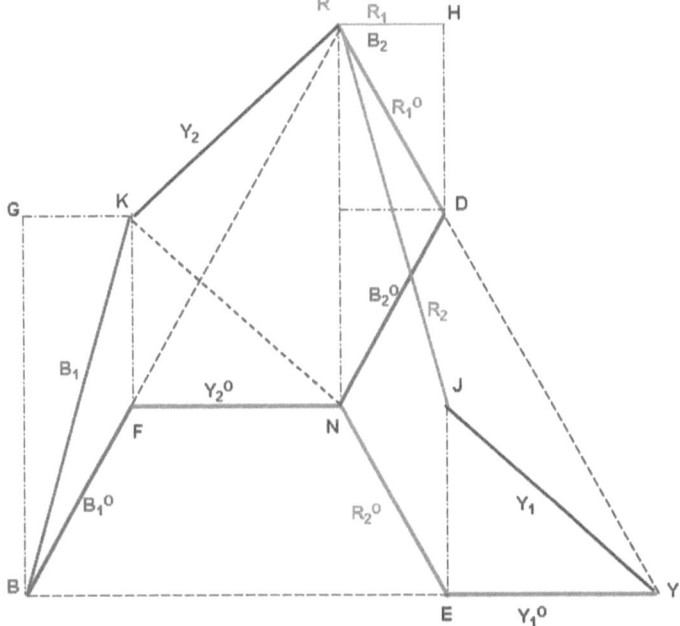

Figure 5-6.4: Phasor Diagram

Analysis of (L-L) Fault on Tertiary Winding in Autotransformer

Appendix 5-7

Rating: 315MVA, 400/220/33kV Autotransformer

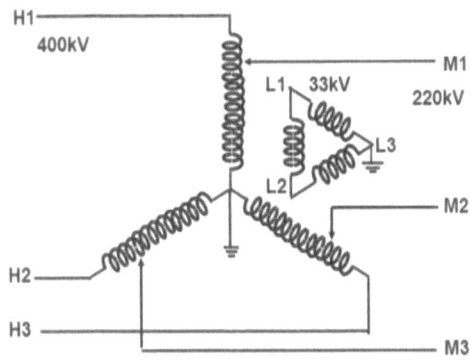

Figure 5-7.1

Data on 315 MVA base
$X_{HM} = 0.125 \; pu$
$X_{HL} = 0.6 \; pu$
$X_{ML} = 0.45 \; pu$

Equivalent Star impedance

$X_H = \dfrac{1}{2}(X_{HM} + X_{HL} - X_{ML}) = 0.1375$

$X_M = \dfrac{1}{2}(X_{HM} + X_{ML} - X_{HL}) = -0.0125$

$X_L = \dfrac{1}{2}(X_{HL} + X_{ML} - X_{HM}) = 0.4625$

Star Equivalent Circuit

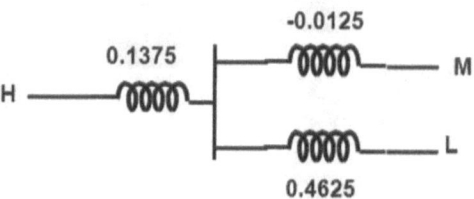

Figure 5-7.2

Assume a (L-G) fault on one corner of delta, Say L1. Since another corner of delta L3 is grounded (Refer Figure 5-7.1), it results in (L-L) fault on delta side.
Assume source fault level of 10,000 MVA on 400KV and 220KV side.

$$X_S \text{ as } 315 \text{ MVA base} = \frac{315}{10,000} = 0.0315 \text{ pu}$$

The positive sequence equivalent circuit can be approximated as given below:

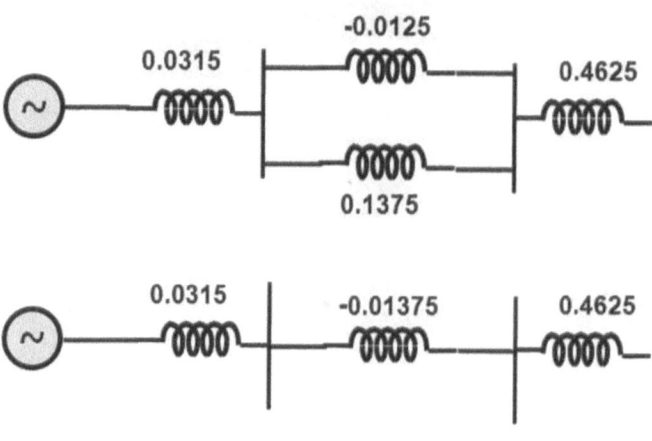

Figure 5-7.3

$Z^+ = 0.48025$

$Z^- = Z^+ = 0.48025$

Analysis of (L-L) Fault on Tertiary Winding in Autotransformer • 193

For (L-L) fault, positive and negative sequence impedances are connected in parallel. The equivalent circuit is given below:

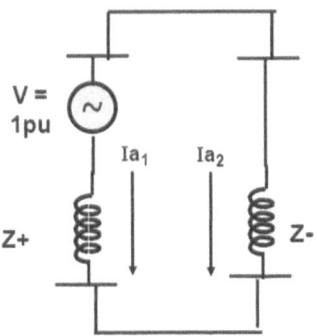

Figure 5-7.4

$$I_a^1 = -I_a^2 = \frac{1.0}{Z^+ + Z^-} = \frac{1.0}{j0.9605} = -j1.0411 \ pu$$

$$I_a^0 = 0$$

$$I_a = I_a^0 + I_a^1 + I_a^2 = 0$$

$$I_b = I_a^0 + a^2 I_a^1 + a I_a^2 \quad (5\text{-}5.1)$$

Where $a = -0.5 + j\sqrt{3}/2$

$\qquad a^2 = -0.5 - j\sqrt{3}/2$

$\qquad 1 + a + a^2 = 0$

From eqn. (5–5.1), $I_b = (a^2 - a) I_a^1$

$\qquad\qquad\qquad = -j\sqrt{3}(-j\,1.0411)$

$\qquad\qquad\qquad = -1.8032 \ pu$

$I_c = I_a^0 + a I_a^1 + a^2 I_a^2$

$\quad = (a - a^2) I_a^1$

$\quad = 1.8032 \ pu$

$1\ p.u.\ current = \dfrac{315}{\sqrt{3} \times 33} = 5.5112 \ KA$

$I_b = -I_c = 1.8032 \times 5.5112$

$\quad = 9.937 \ KA$

The fault will be cleared by unit differential protection within 100 msec. However, for conservative estimate, assume the fault clearance time as 0.5 sec.

As per IS 3043, for aluminium (Table 1.3),

Current rating for 0.5 sec = 178 A/mm²

$$Required\ C/S = \frac{9{,}937}{178} = 56\ sq.\,mm.$$

Choose a cable size of 185 or 240 sq.mm. Since two cables are provided between LV Bushing and grounding grid, adequate margin is considered in sizing.

Suggested CT ratio = 500/1.

Both the cables must pass through the CT ID.

The output of CT shall be connected to over current relay. The suggested time delay is 0.3 to 0.5 sec to allow unit protections to clear the fault within 100 msec.

Grounding – Fairy Tales

Chapter 6

6.0 INTRODUCTION

The flow of ground fault current in electrical power system is intimately related to transformer connections and type of grounding. To evaluate fault current distribution it is necessary to have a basic understanding of transformer theory. Field engineers encounter some 'unexpected or inadvertent' tripping. In this chapter the basic concepts are developed in a series of questions for which answers are given at the end of this chapter. The answers given at the end will give some clue in explaining the 'why' of these events.

QUESTIONS

6.1 Case 1

Figure 6.1 - Fault current returns to (i) A (ii) B (iii) C (iv) D (v) none of the above.
State the reason for your answer

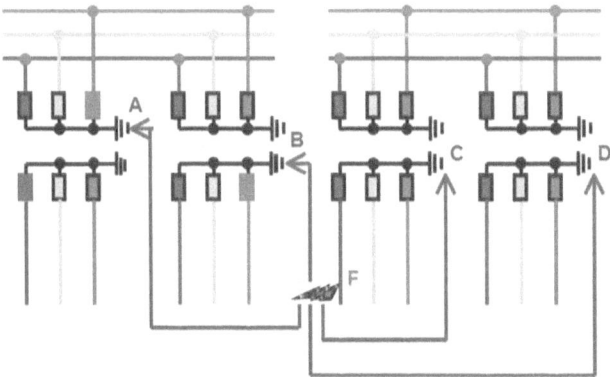

Figure 6.1

6.2 Case 2

Figure 6.2 – Transformer secondary neutral is solidly grounded. Fault current will be high or low?

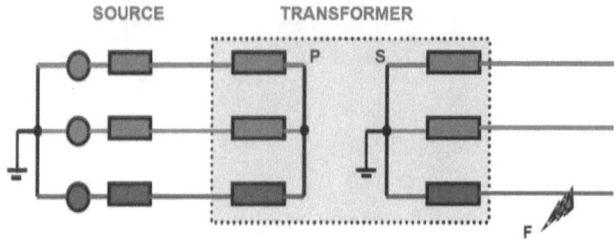

Figure 6.2

6.3 Case 3

Figure 6.3 – Both the primary and secondary neutrals of transformer are solidly grounded. Fault current will be high or low?

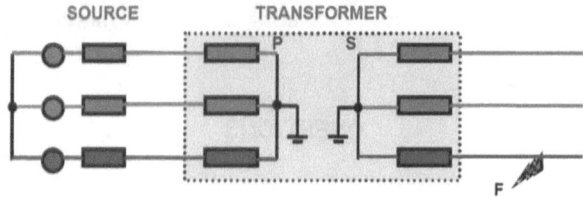

Figure 6.3

6.4 Case 4

Figure 6.4 – Both the primary and secondary neutrals of transformer as well as source neutral are solidly grounded. Fault current will be high or low? Show current distribution in all three phases.

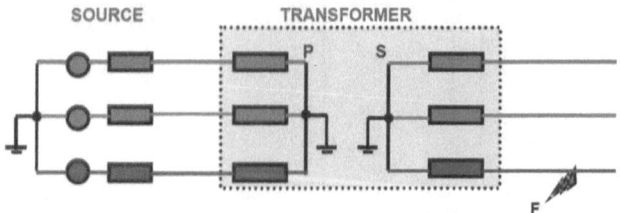

Figure 6.4

6.5 Case 5

Figure 6.5 – Transformer: 11kV/6.6kV; $R_1 = 38\Omega$. Find the current distribution in primary and secondary side.

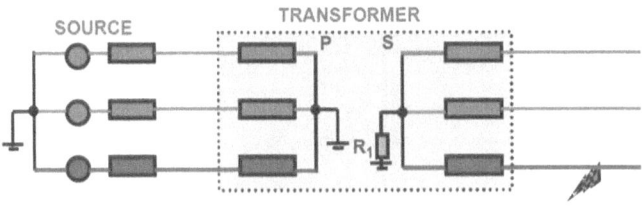

Figure 6.5

6.6 Case 6

Figure 6.6 – Transformer: 11kV/6.6kV; $R_1 = 38\Omega$; $R_2 = 64\Omega$. Find the current distribution in primary and secondary side.

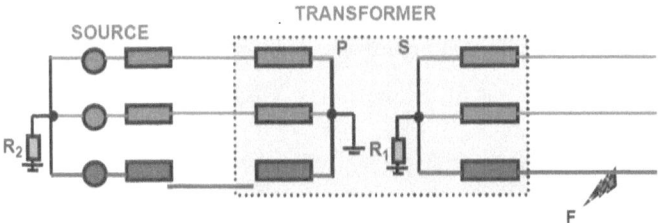

Figure 6.6

6.7 Case 7

Figure 6.7 – Delta – Star Transformer. Show the current distribution in primary and secondary side.

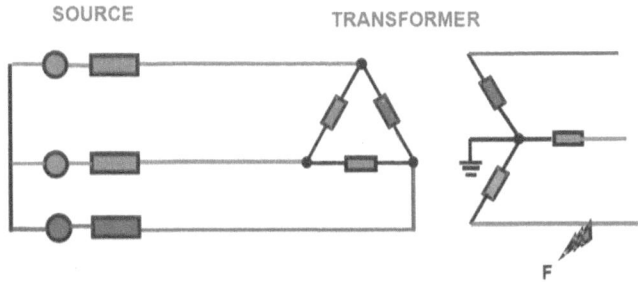

Figure 6.7

6.8 Case 8

Figure 6.8 – Star – Zig Zag Transformer. Show current distribution in primary and secondary side.

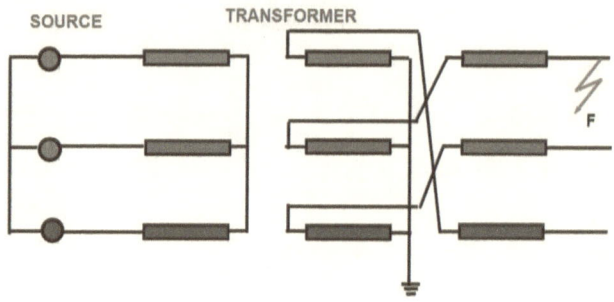

Figure 6.8

6.9 Case 9

Figure 6.9 – Star-Star-Delta tertiary. Source is ungrounded. Can the fault current flow? If so, show the current distribution in primary, secondary side and tertiary.

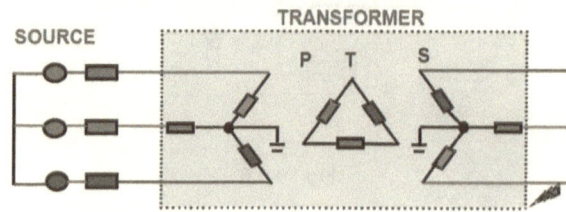

Figure 6.9

6.10 Case 10

Figure 6.10 – Star-Star-Delta tertiary. Source is grounded. Show the current distribution in primary, secondary and tertiary.

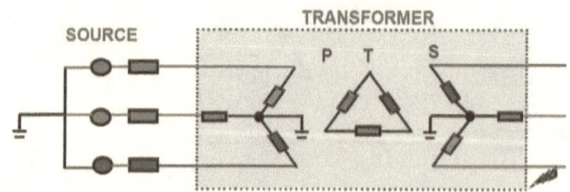

Figure 6.10

6.11 Case 11

Figure 6.11 – Three winding transformer (Star – Star – tertiary) is fed by 220kV system through Over Head line. On 11kV side there is no source. Relay located at R normally responds to fault on downstream side (towards transformer). Can it sense for any ground fault ahead of relay, e.g. for ground fault on EHV line, can the relay at R respond?

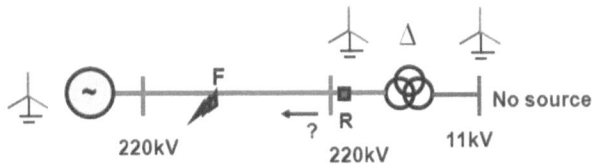

Figure 6.11

6.12 Case 12

Figure 6.12 – Same as Case 11 except the transformer is now conventional Star – Delta transformer. Will the relay located at R respond for upstream ground faults on EHV line?

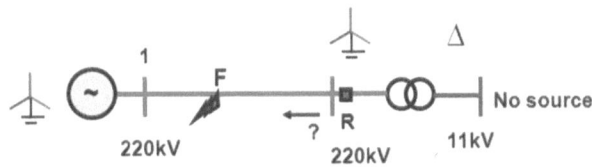

Figure 6.12

6.13 Case 13

Figure 6.13A – Data for analysis:
   Generator at 11kV: 18.75 MVA, $X_1 = X_2 = 20\%$; $X_0 = 10\%$
   Transformer: Star – Delta, 11/33kV, 20MVA, $X_1 = X_2 = X_0 = 8\%$
   11kV Feeder Impedance: Negligible
   No Source is connected on 33kV
   Draw three phase current distribution (Figure 6.13B) for ground fault at F.

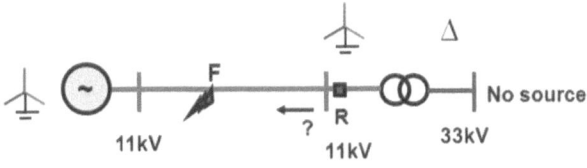

Figure 6.13A

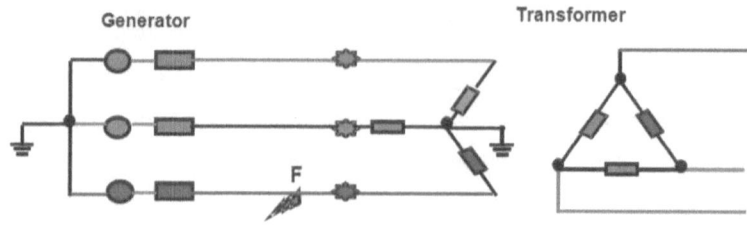

Figure 6.13B

6.14 Case 14

Figure 6.14 – Same as case 13 but with 16Ω resistor in generator neutral and transformer neutral. Draw current distribution for ground fault at F.

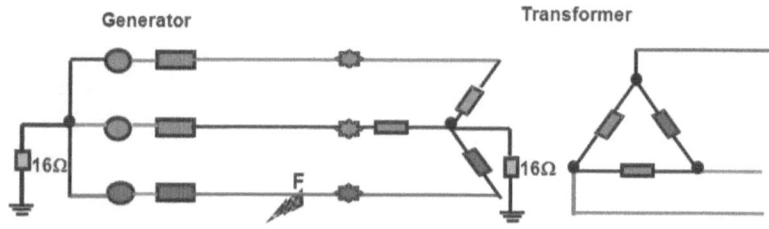

Figure 6.14

6.15 Case 15

SLD of network is shown in Figure 6.15. Generating plant with power evacuation at 400kV is envisaged. 400kV bus is connected to existing 220kV system through 220/400kV ICT. Initially 15.75/400kV GSU (Generator Step Up) transformer is commissioned and back charged from 400kV side. Generator is yet to be commissioned. There was a ground fault on 400kV bus under this operating condition. The GSU tripped on differential protection even though there was no source on LV side.

Find the fault current distribution and justify the reason for pick up of differential relay. This can be considered as a case study to illustrate 'inadvertent grounding'.

Data for analysis:
A) 220kV System Fault Level = 10,000MVA
B) ICT 400/220/33kV
    Impedance on 315MVA Base
    HV – IV = 12.22%
    HV – LV = 61.19%
    IV – LV = 47.55%

C) GSU: 335MVA, 15.75/400kV, 15%

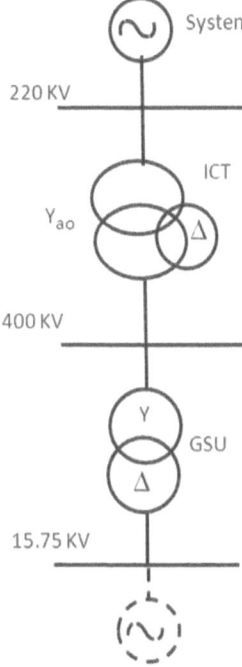

Figure 6.15

6.16 Case 16

A source feeds a transformer. Neutrals of both are solidly grounded. For ground fault on secondary side of transformer, current through transformer neutral shall always be (nearly) equal to ground fault current – True or false?

6.17 Case 17

Can the transformer neutral carry large current *without* ground fault?

ANSWERS

6.0A Answers

Answers to above questions are given below. Readers are encouraged to workout *manually* by drawing equivalent circuits to get a feel of the subject. Then it can be cross checked using software. Majority of software give only terminal currents. If winding current (for example current circulating in delta) is required, special software like PSCAD may have to be used. Supplementary materials wherever required are added to relate with other subjects in power engineering.

6.1A Case 1

(iii) C

Current can't return to neutral of any side of any transformer arbitrarily. The fundamental principle is that the ground fault current will return to that transformer neutral only where AT balance can be achieved. Even in case of so called 'inadvertent grounding', this principle can't be violated.

Concept of AT (Ampere-Turn) balance was already introduced in Sec 5.11.1.1. We will revisit the topic again. Refer Figure 6.16. Windings on the same magnetic limb are shown in similar position. For transformer operation,

$$I_{PRI} \times T_{PRI} = I_{SEC} \times T_{SEC}$$

If this relationship can't be satisfied due to physical connections, no current will flow.

$$I_A \times T_A = I_a \times T_a; \; I_B \times T_B = I_b \times T_b; \; I_C \times T_C = I_c \times T_c$$

The above relationship is almost satisfied in practice except for error introduced by magnetising current (no load) current which is less than 0.5% for modern power transformers.

As per KCL, current returning to neutral is given by

$$I_N = I_A + I_B + I_C$$

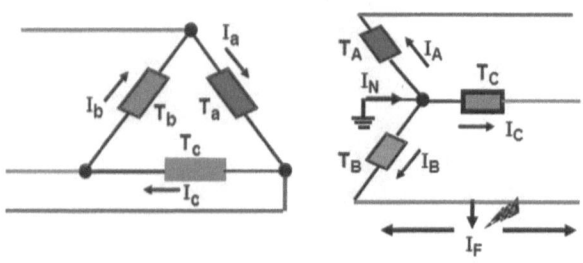

Figure 6.16: AT Balance

6.2A Case 2

No fault current flows. Behaves like ungrounded system

Figure 6.17. Assume prefault current is zero. R and Y phase secondary currents are zero. The corresponding phase currents on primary side are forced to zero to maintain AT balance. If a fault current I_F flows on B phase secondary side, the corresponding current must flow on primary side. But this current can't flow as primary side neutral is open and there is no return path for current. The result is no fault current flows.

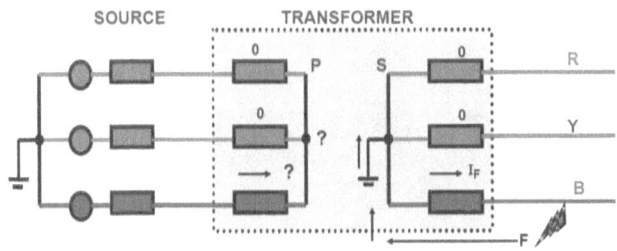

Figure 6.17

Of course purists will argue that this is strictly not true due to 'tank effect'. But these are nuances which can be ignored without affecting the spirit of argument. Zero sequence flux path, zero sequence magnetising reactance and tank effect in three limb core type construction will be covered in later Chapter on Transformer.

6.3A Case 3

No fault current flows. Behaves like ungrounded system

Figure 6.18. Same as case 2 above except in this case source neutral is open and no return path is available for completing the circuit on primary side. Even though neutrals of both primary and secondary side are grounded, still there is no current flow.

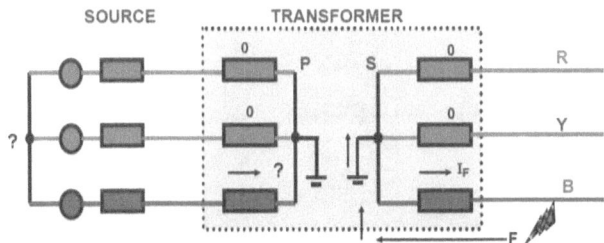

Figure 6.18

6.4A Case 4

Large fault current flows. Behaves like truly grounded system Figure 6.19.

Both primary and secondary neutrals grounded and source also grounded. Currents can flow in both the secondary and primary sides of transformer and satisfy AT balance and KCL.

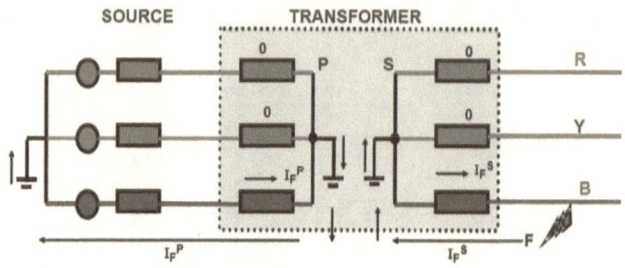

Figure 6.19

6.5A Case 5

Figure 6.20.

- Transformer secondary neutral is grounded through resistance.
- Transformer primary neutral and source neutral are solidly grounded
- Currents can flow in both the secondary and primary sides of transformer
- Ground fault current on secondary side is limited by NGR value
- Only the equivalent reflected current of reduced secondary current flows on the primary side, even though primary neutral and source neutral are solidly grounded
- Example: Transformer 11KV/6.6KV, R_1 = 38Ω
- Ground fault current on secondary side = $(6600/\sqrt{3})/38$ = 100 A
- Equivalent current on primary side = 100 * 6.6/11 = 60A

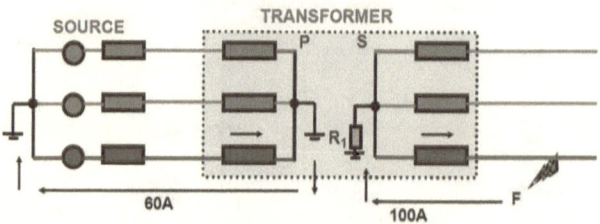

Figure 6.20

6.6A Case 6

Figure 6.21.
- Transformer secondary neutral and source grounded through resistance
- Transformer primary neutral solidly grounded
- Currents can flow in both the secondary and primary sides of transformer
- Ground fault current on secondary side limited by NGR (R_1) value
- The equivalent reflected current on the primary side encounters another NGR (R_2) on source side.
- Overall effect: Fault current reduces substantially
- Example: Transformer 11KV/6.6KV, $R_1 = 38\Omega$, $R_2 = 64\Omega$
- Turns Ratio TR = 6.6/11 = 0.6
- R_2 reflected on secondary side (Refer Sec 5.9.1.1) = $R_2' = TR^2 \times 64 = 23$
- Total Resistance on neutral of secondary = $R_1 + R_2' = 61\Omega$
- Ground fault current on secondary side (F_1) = $(6600/\sqrt{3})/61 = 62$ A
- Equivalent current on primary side = 62 * 6.6/11 = 37A
- Consider ground fault on primary side (F_2).

Ground fault current = $(11000/\sqrt{3})/64 = 100$A

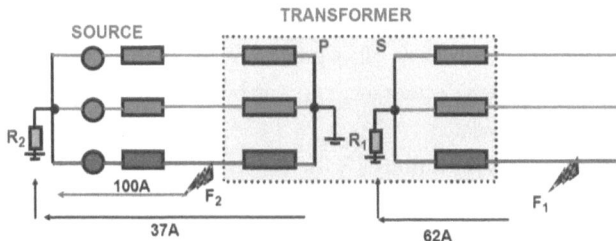

Figure 6.21

6.7A Case 7

Figure 6.22.

Transformer vector group is Delta - Star Grounded. Transformer is assumed to be unloaded. Y and B phase secondary currents are zero. The corresponding winding currents on primary side are forced to zero to maintain AT balance. If a fault current I_F flows on R phase secondary side, the corresponding winding current must flow on primary side. Since the currents on other two windings are forced to zero, the reflected fault current on primary side is forced as (L-L) fault.

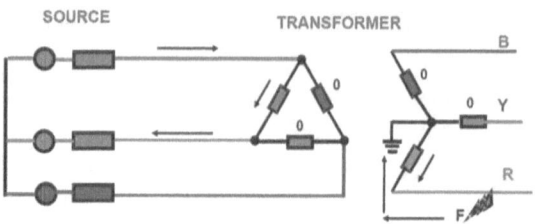

Figure 6.22

Thus, in case of delta –star transformer, line to ground fault on star side is reflected line to line fault on delta side. Ground fault isolation (also termed as zero sequence isolation) is 'naturally' obtained in delta – star transformation. This has profound impact on protection.

Relays responding to ground fault on star side are
(i) Phase relay on faulted phase (51R)
(ii) Ground relay (51N)

On delta side, phase relays on two phases (51R, 51Y) will respond. More importantly, Ground relay does not pick up!

Settings of phase relays (51) are typically in the range of 125% to 200% while the ground relay setting is 10% to 40%. Since ground relay (51N) on delta side will not pick up for ground faults on star side, nuisance tripping by sensitive ground relays on delta side for faults on star side is obviated.

Another interesting observation is with regard to voltages on star and delta side (Figure 6.23). The star side voltage will be similar to Figure 4.7 as per solidly grounded system. On delta side, the line voltage triangle is almost isosceles. One phase voltage V_{YG} is near normal but other two phase voltages (V_{RG} & V_{BG}) are almost half of normal value. But open delta voltage ($V_{RG}+V_{YG}+V_{BG} = 3V_0$) is zero! Of course this is expected as the reflected fault current on delta side does not involve ground.

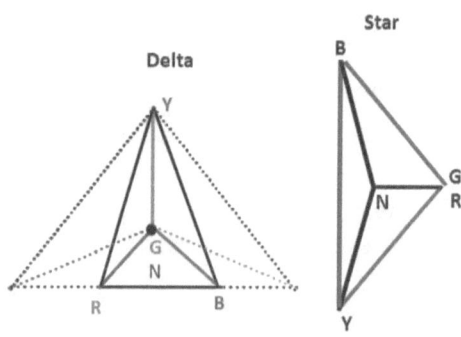

Figure 6.23: Phasor Diagram

6.8A Case 8

Figure 6.24.

Since transformer is unloaded, current through windings ZigY and ZigB are zero. Hence the current through series connected windings ZagB and ZagR are forced to zero. Since ZigB and ZagB carry no current, current on star winding on the same magnetic limb (B phase) is forced to zero. The reflected current on star winding is again forced as (L-L) fault as in case 7. Grounding of transformer primary neutral and source neutral is immaterial. Ground fault isolation (zero sequence isolation) between primary and secondary is 'naturally' obtained in this case also.

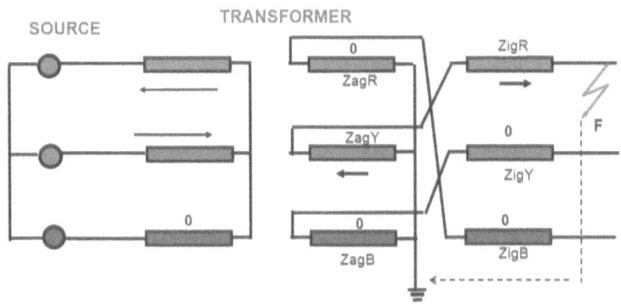

Figure 6.24

6.9A Case 9

Figure 6.25

- Transformer is Star Grounded [P] – Δ[T] – Star Grounded [S]
- Source ungrounded
- Under normal condition Tertiary does not carry any current. Under single phase to ground fault conditions zero sequence currents can circulate within delta. This facilitates substantial ground fault current flow.
- Primary, Secondary and Tertiary windings on the same magnetic limb are shown in similar geometric position.
- Typical current (in pu) distribution is shown for fault on secondary side
- Secondary R phase current is balanced by primary and tertiary R phase currents (3 = 2+1).
- Primary B and Y phase currents are balanced by tertiary B and Y phase currents (=1) with zero secondary currents as transformer is considered unloaded.
- Ampere – turns balance and KCL satisfied.
- No current through primary neutral as source is ungrounded

- Source ungrounded or transformer primary neutral ungrounded will result in same current distribution.
- Reflected fault current on primary side is in the ratio of 2:1:1 without neutral current.

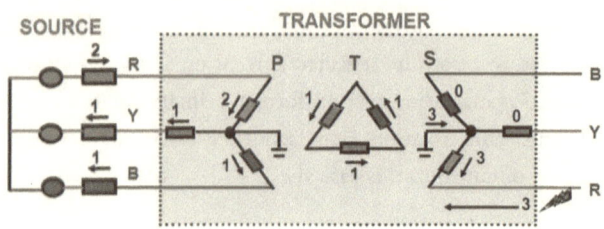

Figure 6.25

6.10A Case 10

Figure 6.26

- Transformer is Star Grounded [P] – Δ[T] – Star Grounded [S]
- Source neutral also grounded
- Typical current (pu) distribution shown for fault on secondary side
- Ampere – Turns balance and KCL satisfied
- Because of presence of Tertiary, current (pu) through Primary neutral is less than current (pu) through Secondary neutral. In fact the difference is contributed by current circulating in Tertiary. Current division between Tertiary and Primary is dependent on zero sequence leakage reactance between Primary to Tertiary and Secondary to Tertiary.

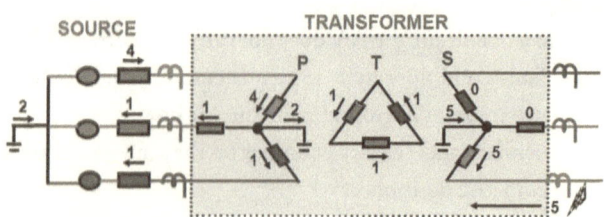

Figure 6.26

Remarks on Differential Protection for Star – Star transformer

Because of delta tertiary, per unit ground fault currents flowing on secondary and primary sides are not equal. Hence, the currents measured by line CTs on secondary and primary sides (accounting for turns ratio) are not equal; spill current flows in operating coil of relay making the differential scheme to pick up for external faults. By providing a star - delta IPCT (Figure

6.27), "through circuit zero sequence unbalance" is prevented from entering the operating coil of relay, ensuring stability for external faults. Note how KCL is satisfied at every node. Thus the presence or absence of tertiary winding has significant impact on protection schemes.

In modern numerical relays IPCTs are software implemented to filter out zero sequence currents. More extensive discussions on transformer differential protection will be covered in later chapters.

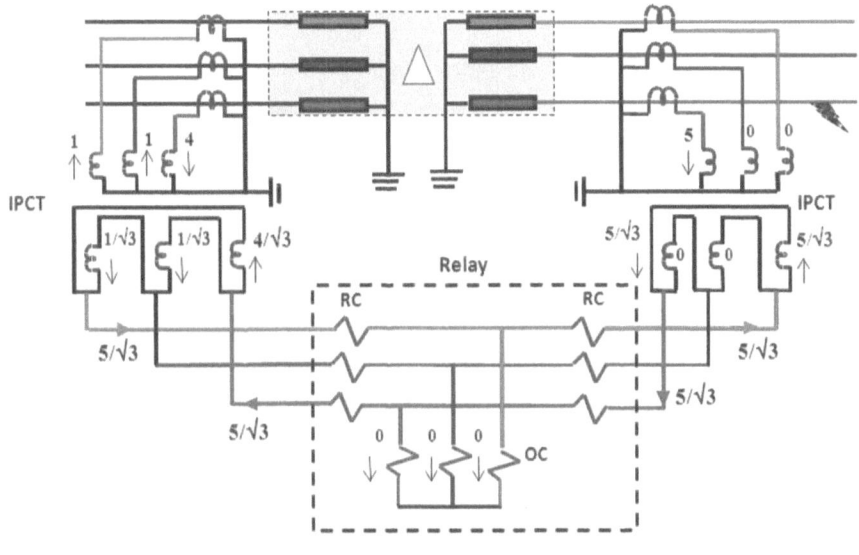

Figure 6.27

6.11A Case 11

Figure 6.28 [2]
- Transformer is Star Grounded [P] – Δ[T] – Star Grounded [S]
- Source neutral also grounded
- Data for sample study:
  - Transformer 220/11/3.3kV, HV - LV - T
  - Impedance on 30 MVA Base
    - HV – LV: 12.1%
    - LV – T: 21.2%
    - HV – T: 36.3%
    - $X_1 = X_2 = X_0$
- 220kV O/H Line data on 10MVA base:
  - $Z_1 = Z_2 = 0.00008 + j\,0.00045\,\text{pu}$
  - $Z_0 = 0.00033 + j\,0.00115\,\text{pu}$

- 220kV System Fault level ($F_{3PH} = F_{1PH}$) = 2983MVA
- Current distribution (in pu on 10 MVA Base) for fault on incoming line is shown in Figure 6.28. Majority of ground fault current returns to system neutral. The other part returns through HV side neutral of Transformer. The balancing zero sequence currents flow within delta tertiary.

- I_{BASE} on 220kV = $\dfrac{10}{(\sqrt{3} \times 220)}$

  = 26.25A

- Fault current I_F = 301 x 26.25 = 7.9kA
- Current sensed by ground fault relay (51N) at location R:

  $3I_0$ = 8.1pu = 8.1 x 26.25 = 213A

  Normally we expect the relay to sense for faults on downstream side. In this case relay senses for faults ahead of transformer and if setting is too low it can trip the transformer also. If the transformer also trips along with the line, fault location becomes more tedious. To overcome the above, it is a normal practice to directionalise the ground relay R(67N) so that it looks only for faults within the transformer and 11kV bus. Needless to add that this will require voltage polarization signal (V_0) from open delta PT output. Refer Cl 9.17 of [21] for further details on directional earth fault protection.

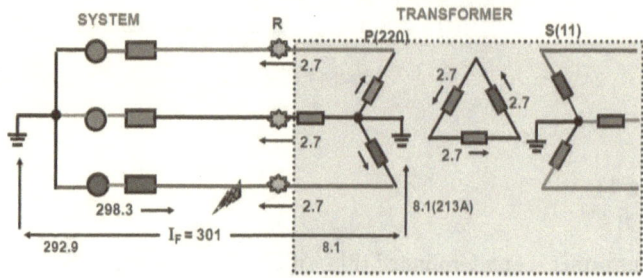

Figure 6.28

6.12A Case 12

Figure 6.29

- Same as Case 11 except the transformer is conventional two winding star – delta transformer with reactance of 12.1% on 30MVA base. In case 11, HV-T reactance was much higher (36.3%). Hence in this case we can anticipate higher ground fault current to return via transformer neutral.
- Current distribution (in pu on 10 MVA Base) for fault on incoming line is shown in Figure 6.29.

- Fault current $I_F = 306 \times 26.25 = 8kA$
- Current sensed by ground fault relay (51N) at location R:
$3I_0 = 22.8pu = 22.8 \times 26.25 = 599A$
This is almost 3 times that in Case 11. More chance in this case for ground fault relay to pick up. This is a case of 'inadvertent grounding'.

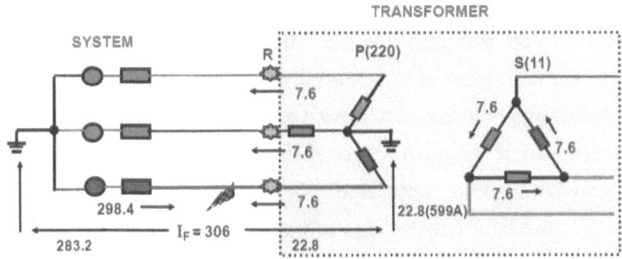

Figure 6.29

6.13A Case 13
Figure 6.30
- Very similar to Case 12, except the system is isolated generator.
- Generator: 18.75 MVA, $X'_d = X_1 = X_2 = 20\%$, $X_0 = 10\%$
- Conventional two winding star – delta transformer, 11/33kV, 20MVA, $X_1 = X_2 = X_0 = 8\%$
- Both generator and transformer solidly grounded.
- Current distribution (in pu on 10 MVA Base) for fault on incoming line is shown in Figure 6.30.
- I_{BASE} on 11kV = $10/(\sqrt{3} \times 11) = 525A$
- Fault current $I_F = 12.7pu = 12.7 \times 525 = 6.67kA$
- Current through generator neutral = $5.5pu = 5.5 \times 525 \cong 2.89kA$.
- Current through transformer neutral = $7.2pu = 7.2 \times 525 \cong 3.78kA$.
The balancing zero sequence currents flow within secondary delta.
- As before, The relay located at R can sense for faults ahead of transformer

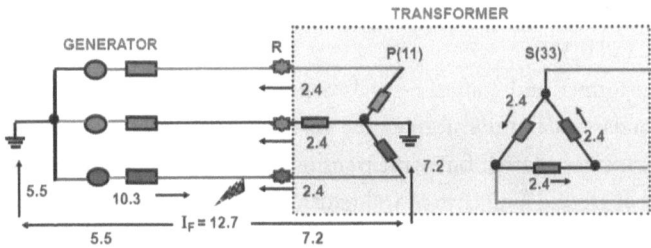

Figure 6.30

6.14A Case 14

Figure 6.31

- Both generator and transformer grounded through resistance of 16Ω. Expected current through NGR = (11000 /√3)/16 = 400A
- Current distribution (in pu on 10 MVA Base) for fault on incoming line is shown in Figure 6.31. I_{BASE} = 10/(√3 x 11) = 525A
- Current through generator neutral = 0.75pu = 0.75 x 525 = 383A
- Current through transformer neutral = 0.75pu = 0.75 x 525 = 383A
- The above values closely match with expected values.
- The fault current is almost resistive while in case 13 it is almost reactive.
- It is interesting to compare Figure 6.21 with Figure 6.31. In Figure 6.21, for fault at F_2, the fault current is purely dependent on value of R_2. In Figure 6.31, the fault current is dependent on values of R_1 and R_2. This is due to difference in vector group of transformer (Star-Star vs Star – Delta).
- Again, the relay located at R can sense for faults ahead of transformer
- Cases 12 to 14 bring out an important fact that relay setting and coordination shall be done with caution to avoid inadvertent tripping.

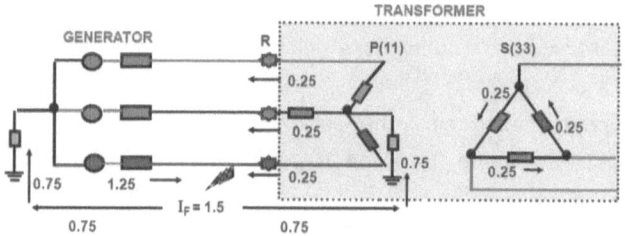

Figure 6.31

6.15A Case 15

Refer Appendix 6-1 for analysis.

6.16A Case 16

Figure 6.32A and 6.32B

- Transformer and source neutrals are solidly grounded. For ground fault on secondary side of transformer, the ground fault current is expected to return to transformer neutral. But if the transformer is *autotransformer with delta tertiary*, part of ground fault current will return to transformer neutral and part return to source neutral. Under some conditions, entire ground fault current can return to source neutral bypassing transformer neutral.

- Data for simulation (Figure 6.32A):
- Autotransformer: 138(M)/230(H)/11(L) kV, Y-Y-Δ
  On 37.5MVA Base,
  X_{HM} = 9.5%; X_{ML} = 47.38%; X_{HL} = 31.02%
  Source: $X_1 = X_2$ = 20%; X_0 = 10%
  The above example is taken from Sec 9.10 of [19] but the impedance data is slightly modified.

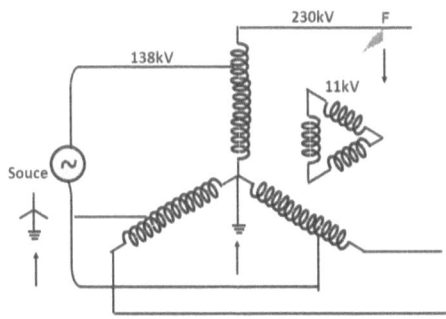

Figure 6.32A

- For ground fault at F, current distribution in Amp is indicated in Figure 6.32B. Since it is an autotransformer, primary and secondary are electrically connected and KCL has to be satisfied at all nodes. The distribution is so unique that current through transformer neutral is zero even though it is solidly grounded. The fault current returns to source neutral directly bypassing the transformer! If standby earthfault relays (51SN) are provided on transformer neutral and source neutral, relay on source neutral will pick up and trip the source.

This may mislead to interpret that the fault location is between source and transformer.

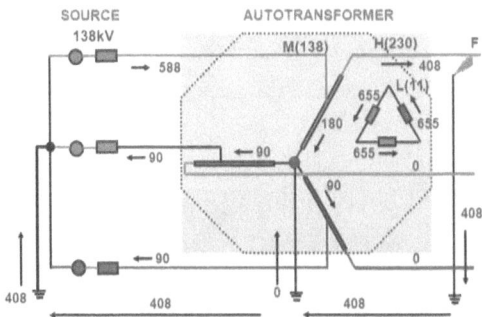

Figure 6.32B

6.17A Case 17

Figure 6.33A and 6.33B

- ▷ Similar to example considered in Sec 7.4.1 of [19] but data modified.
- ▷ Can ground current (zero sequence) flow without ground fault?
  Of course yes!
  It is illustrated with an example. Data for simulation:
  Transformer: 33/11kV, YNd, 30MVA, 10%
  Load: 28.6MW at UPF
  Under normal operating condition, the load current is 500A on 33kV side and neutral current is zero. Refer Figure 6.33A.

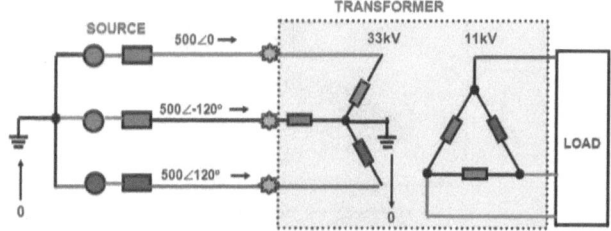

Figure 6.33A

- ▷ Current distribution with R phase on 33kV side open (e.g. conductor snapping without touching ground) is shown in Figure 6.33B.
- ▷ Ground current (zero sequence current) = 1446A
- ▷ Without ground fault, large ground fault current flows on primary side.
- ▷ Earth fault relays (51N) will pick up. The operator may be misled that a shunt fault to ground has occurred while it is a series (open conductor) fault.
- ▷ In EHV system, 3 single phase breakers, one for each phase is provided. During closing operation, assume two poles close and one pole is open. This is similar to the case in Figure 6.33B. Pole Discrepancy detection also uses earth fault relays.
- ▷ This case is converse of case 16. In this case, there is a large neutral current flow in transformer without ground fault. In case 16, neutral current flow in autotransformer is zero even in the presence of ground fault.

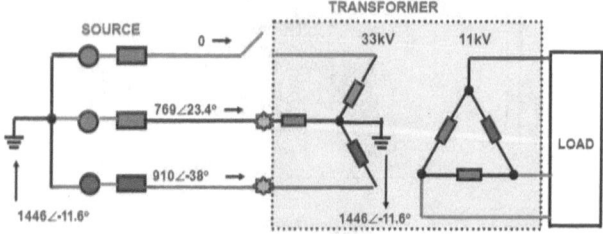

Figure 6.33B

Analysis of Tripping of GSU By Differential Protection for External Fault

Appendix 6-1

6-1.0 INTRODUCTION

The following analysis is done to estimate the current in various branches for a (L-G) fault on 400 KV side bus with unloaded GSU (Generator Step Up Transformer) connected to 400 KV side.

6-1.1 DATA
A) SLD

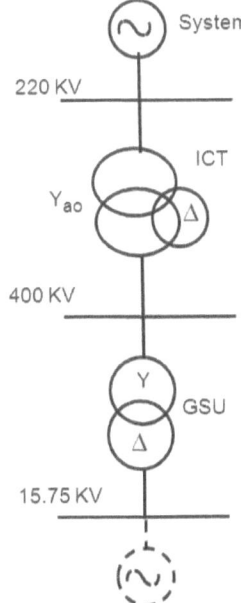

Figure 6-1.1

B) Base MVA = 315

C) 220kV System Fault Level = 10,000MVA

$$\text{System } X_{sys} = \frac{\text{Base MVA}}{\text{Fault Level}}$$

$$= \frac{315}{10,000} = 0.0315 \text{pu}$$

D) ICT 400 / 220 / 33kV
Impedance on 315MVA Base
HV – IV = 12.22%
HV – LV = 61.19%
IV – LV = 47.55%

E) GSU: 335MVA, 15.75/400kV, 15%

6-1.2 ANALYSIS
Base MVA = 315

6-1.2.1 ICT Star Equivalent:

$$(400\text{kV}) \text{ HV–N} = \frac{12.22 + 61.19 - 47.55}{2} = 12.93\%$$

$$(220\text{kV}) \text{ IV–N} = \frac{12.22 + 47.55 - 61.19}{2} = -0.71\%$$

$$(33\text{kV}) \text{ LV – N} = \frac{61.19 + 47.55 - 12.22}{2} = 48.26\%$$

Star Equivalent circuit for three winding transformer

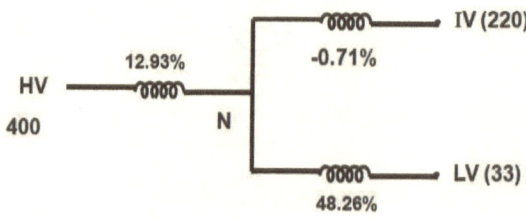

6-1.2.2 GSU

Z = 15% on 335MVA.

On 315 MVA base, $Z_1 = Z_2 = Z_0 = 15 \times \dfrac{315}{335}$

$= 14.1\%$

For (L-G) a fault on 400 KV side, this acts like grounding transformer when there is no source on LV side.

6-1.2.3 Base Current

At 400KV, $I_{BASE} = \dfrac{315}{\sqrt{3} \times 400} = 0.4547$ KA

At 220KV, $I_{BASE} = \dfrac{315}{\sqrt{3} \times 220} = 0.8267$ KA

6-1.2.4 Sequence Network for fault on 400 KV Side.

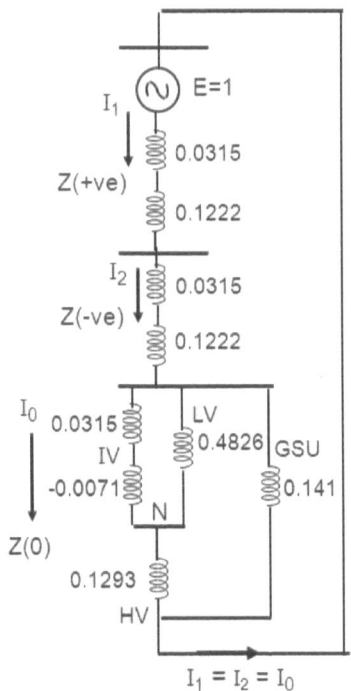

Figure 6-1.2

$Z_1 = 0.0315 + 0.1222 = 0.1537$pu

$Z_2 = 0.1537$pu

Zero sequence impedance is obtained by network reduction:

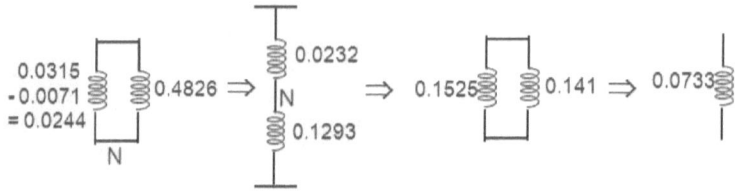

Figure 6-1.3

$Z_0 = 0.0733\text{pu}$

$I_1 = I_2 = I_0 = \dfrac{1}{Z_1+Z_2+Z_0}$

$= \dfrac{1}{0.1537+0.1537+0.0733}$

$= \dfrac{1}{0.3807}$

$= 2.6267\text{pu}$

6-1.2.5 Fault Current

$I_F = 3I_0 = 7.8802\text{ pu}$
$I_F = \text{Fault Current} = 7.8802 \times I_{BASE}$
$\qquad\qquad\qquad\quad = 7.8802 \times 0.4547$
$\qquad\qquad\qquad\quad = 3.5831\text{kA}$

6-1.2.6 Contribution from 400 KV side to Fault (no source on 15.75kV side)

$I_1 = I_2 = 0$
$I_0 \text{ (From GSU)} = \dfrac{0.1525}{(0.1525+0.141)} \times 2.6267$

$\qquad\qquad = 0.5196 \times 2.6267$
$\qquad\qquad = 1.3648\text{pu}$
$\qquad\qquad = 1.3648 \times I_{BASE}$
$\qquad\qquad = 1.3648 \times 0.4547$
$\qquad\qquad = 0.6205\text{kA}$

$I_a = I_b = I_c = I_0 = 0.6205\text{kA}$

6-1.2.7 Circulating current in GSU secondary (Δ) winding

$$\frac{0.6205 \times \frac{400}{\sqrt{3}}}{15.75}$$

= 9.1kA

6-1.2.8 Contribution from 220 KV side

$I_1 = I_2 = 2.6267$ pu
I_0 = Contribution from 220KV Side (Figure 6-1.4)

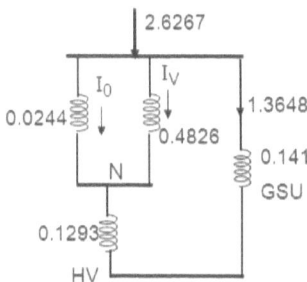

Figure 6-1.4

$$I_0 = \frac{0.4826}{(0.0244 + 0.4826)}(2.6267 - 1.3648)$$

= 0.9519 x 1.2619
= 1.2012pu

$I_a = I_0 + I_1 + I_2$
= 1.2012 + 2.6267 + 2.6267
= 6.4546pu
= 6.4546 x I_{BASE}
= 6.4546 x 0.8267kA
= 5.3361kA

$I_1 = I_2$
$I_b = I_0 + a^2 I_1 + a I_2$
= $I_0 + (a^2 + a) I_1$
= $I_0 - I_1$
= 1.0212 − 2.6267
= − 1.4255pu.
= − 1.4255 x 0.8267kA
= − 1.1785kA

$I_c = I_0 + aI_1 + a^2 I_2$
$ = I_0 + (a^2 + a) I_1$
$ = I_b$
$ = -1.1785\text{kA}$

6-1.2.9 Current in ICT Tertiary using (AT) Balance.

Phase RY

$$\frac{\left(2.3735 \times \frac{220}{\sqrt{3}}\right) - \left(2.9626 \times \frac{400-220}{\sqrt{3}}\right)}{33} = 0.19$$

$I_0^{RY} = 0.19\text{kA}$

Phase YB, BR

$$\frac{\left(0.558 \times \frac{220}{\sqrt{3}}\right) - \left(0.6205 \times \frac{400-220}{\sqrt{3}}\right)}{33} = 0.19$$

$I_0^{YB} = I_0^{BR} = I_0^{RY} = 0.19\text{kA}$

6-1.2.10 Current Distribution

Current Distribution (in kA) is shown in Figure 6-1.5. The reader is encouraged to check how KCL is satisfied at every node. As a final check,

$\Sigma I_{OUT} = 3.5831 + 1.2575 = 4.8406\text{pu}$
$\Sigma I_{IN} = 1.8615 + 2.9791 = 4.8406\text{pu}$

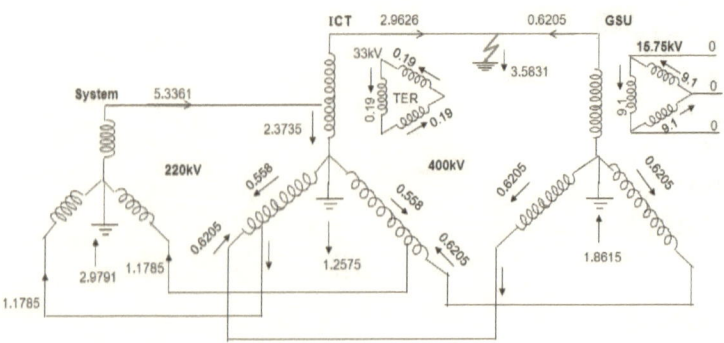

Figure 6-1.5

6-1.2.11 GSU Differential Relay Pick Up

Line current on LV side of GSU is zero (current circulates within delta only). Line current on 400kV side is 620A. The differential for GSU picks up even though the fault is not within the transformer.

Generator Neutral Grounding Practices

Chapter 7

7.0 INTRODUCTION

Various methods of grounding are described with their impact on zero sequence current circulation and core damage in rotating equipment [7]. The grounding methods for LV and MV generators are elaborated. The principles of high and low resistance grounding are discussed leading to sizing of NGT and NGR. The dangers of mixing up incompatible grounding systems are brought out.

7.1 HARMONIC AND ZERO SEQUENCE

The theory of symmetrical components defines three components for three phase system:

I_{POS}: Positive sequence – (e.g. $1\angle 0°$, $1\angle -120°$, $1\angle -240°$)
I_{NEG}: Negative sequence – (e.g. $1\angle 0°$, $1\angle -240°$, $1\angle -120°$)
I_{ZER}: Zero sequence – (e.g. $1\angle 0°$, $1\angle 0°$, $1\angle 0°$)

The relationship between harmonics and sequence component can be derived from Table 7-1 [2].

Harmonics Number	Phase Angle of Harmonic Components			Phase Rotation
	Phase R	Phase Y	Phase B	
Fundamental (1)	0	120	240	(+ve)
Second (2)	0	2 x 120(=240)	2x240(=120)	(-ve)
Third (3)	0	3 x 120(=0)	3x240(=0)	Zero
Fourth (4)	0	4 x 120(=120)	4x240(=240)	(+ve)
Fifth (5)	0	5 x 120(=240)	5x240(=120)	(-ve)
Sixth (6)	0	6 x 120(=0)	6x240(=0)	Zero

Contd...

Phase Angle of Harmonic Components				
Harmonics Number	Phase R	Phase Y	Phase B	Phase Rotation
Seventh (7)	0	7 x 120(=120)	7x240(=240)	(+ve)
Eighth (8)	0	8 x 120(=240)	8x240(=120)	(-ve)
Ninth (9)	0	9x120(=0)	6x240(=0)	Zero
Tenth (10)	0	10 x 120(=120)	10x240(=240)	(+ve)
Eleventh (11)	0	11 x 120(=240)	11x240(=120)	(-ve)
Twelfth (12)	0	12x120(=0)	12x240(=0)	Zero
Thirteenth (13)	0	13 x 120(=120)	13x240(=240)	(+ve)
Fourteenth (14)	0	14 x 120(=240)	14x240(=120)	(-ve)
Fifteenth (15)	0	15 x 120(=0)	15x240(=0)	Zero

Table 7.1

An easier way to recall sequence component for a particular harmonic is given in Table 7.2.

Positive	Negative	Zero
1	2	3
4	5	6
7	8	9
10	11	12
13	14	15

Table 7.2

For example, 5th harmonic is negative sequence and 10th harmonic (if even harmonics are present) is positive sequence. Multiples of 3rd harmonic (3, 6, 9, 12, 15) are zero sequence. Refer Figure 7.1. 3rd harmonic is a zero sequence quantity at 150 Hz, 9th harmonic is a zero sequence quantity at 450 Hz, and so on. Zero sequence current used in fault current calculations is at 50Hz.

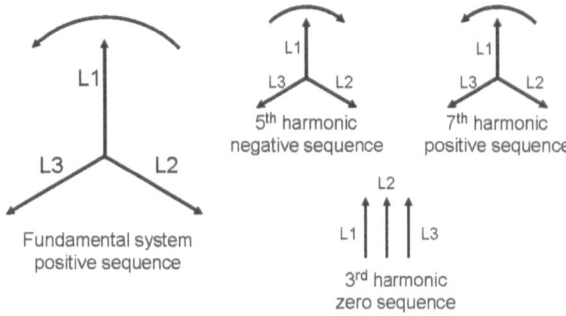

Figure 7.1: Harmonics and Sequence Components

In case of positive (or negative) sequence quantities, neutral grounding is immaterial as neutral current is zero. This is precisely the reason why the classical conversions of star to delta and delta to star taught in first course in electrical engineering are possible for positive (or negative) sequence quantities.

But for flow of zero sequence current, neutral connection to 'ground' must exist as shown in Figure 7.2. Hence, power engineers use the terms 'zero sequence current' and 'ground fault current' interchangeably. If neutral connection does not exist, as in ungrounded system, suppression of zero sequence current appears as zero sequence voltage. Open delta PT (Refer Sec 4.9) employed in ungrounded system measures this voltage ($3V_0$).

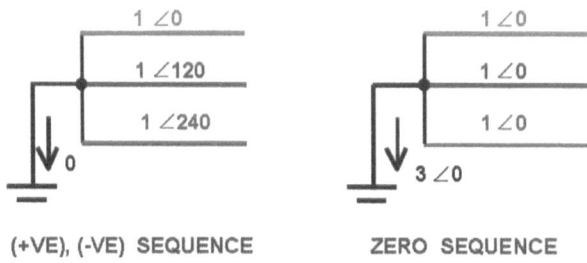

Figure 7.2: Neutral Current

7.2 GENERATORS CONNECTED TO A COMMON BUS

The scheme is shown in Figure 7.3.

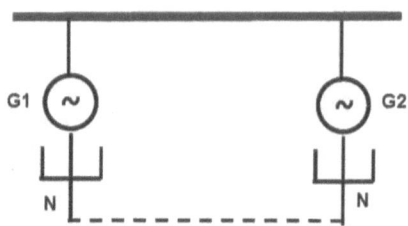

Figure 7.3: Generators connected to common bus

(i) Non-zero-sequence currents

The current flow is shown in Figure 7.4.

Non-zero-sequence currents (1, 5, 7, 11, 13...) flow between the machines *irrespective of neutrals grounded or not*. Fortunately the magnitudes of higher order harmonic (5, 7, 11...) generated in alternators are very less and hence result in insignificant circulating currents (I_5, I_7 ...).

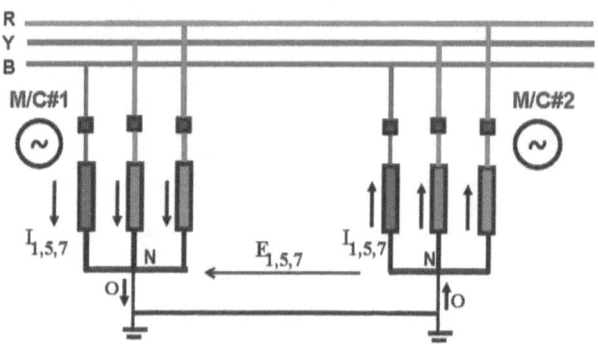

Figure 7.4: Flow of (+ve) or (-ve) sequence currents

In case of fundamental, the difference between internal emf of machines 1 and 2 leads to circulating reactive flow and bus voltage hunting. This typical problem encountered in practice can be mitigated by proper choice of AVR droop setting for respective machines. In practice, only small machines (less than a few MW capacity) operate in parallel connected to a common bus. In this case, AVR droop setting for individual machines can be adjusted to minimize hunting of bus voltage. Definition of AVR droop setting is very similar to governor droop setting of turbine. In Figure 7.5, AVR droop characteristics of two generators rated 1000KVA each operating in parallel are shown. Generator-1 is set for 10% droop and Generator-2 is set for 5% droop. If the common bus voltage is 98%, from the characteristics,

MVAR share from Generator-1 = (1000/10) x 2 = 200KVAR
MVAR share from Generator-2 = (1000/5) x 2 = 400KVAR

By adjusting the droop, KVAR sharing of individual generators can be controlled and is a practical tool to stabilize the bus voltage.

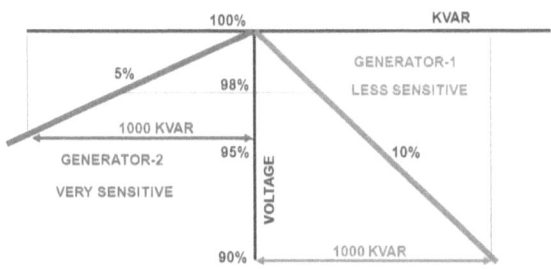

Figure 7.5

(ii) Zero-sequence currents

The alternators generate certain amount of third harmonic voltage. Depending on pitch factor, magnitude of third harmonic voltage generated varies. Under healthy conditions third harmonic voltages are present near to the neutral and terminal. Typically the magnitude of third harmonic voltage is 1% to 3% of generator phase voltage. Any differential third harmonic voltage between machines will result in circulation of third harmonic currents among machines *if the neutrals are directly tied together* as shown in Figure 7.4.

In case of fundamental positive sequence voltage, 1% difference in voltage between internal emfs of two machines may not result in large circulating current.

$X_1 = X_d \cong 200\%$
$\Delta E = 1\%$

$$\text{Circulating Current } I_C = \frac{\Delta E}{(X_1 + X_1)} = \frac{0.01}{(2+2)} = 0.0025 = 0.25\% \, I_{RAT}$$

In case of third harmonic voltage, which is zero sequence, 1% difference in voltage between internal emfs of two machines results in significant circulating current.

$X_0 \cong 10\%$
$\Delta E = 1\%$

$$\text{Circulating Current } I_C = \frac{\Delta E}{(X_0 + X_0)} = \frac{0.01}{(0.1+0.1)} = 0.05 = 5\% \, I_{RAT}$$

7.3 NGR COMMON TO ALL THE UNITS

This method is widely used in DG (Diesel Generator) plants (Figure 7.6).

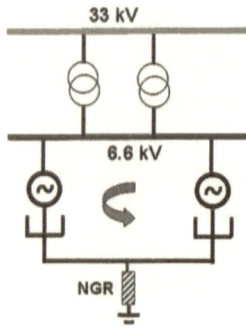

Figure 7.6: Common NGR

The ground fault current supplied by the plant is limited by common Neutral Grounding Resistor (NGR) and remains almost the same irrespective of number of units operating in parallel. This leads to simplified ground fault relaying. In this case ground relays with DMT characteristics are very suitable.

The disadvantage is the circulation of significant zero sequence current among machines as neutrals are connected with low/zero impedance (Figure 7.7).

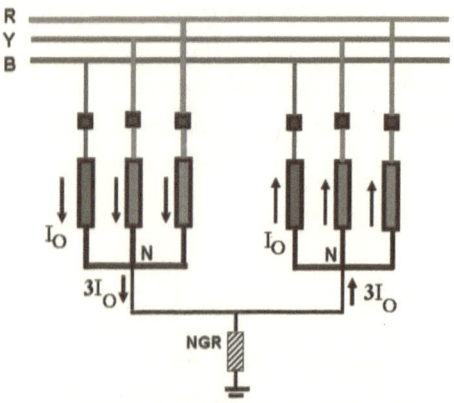

Figure 7.7: Zero Sequence Current Flow

7.4 INDIVIDUAL NGR FOR THE UNITS

(i) Generators on common bus (Figure 7.8)

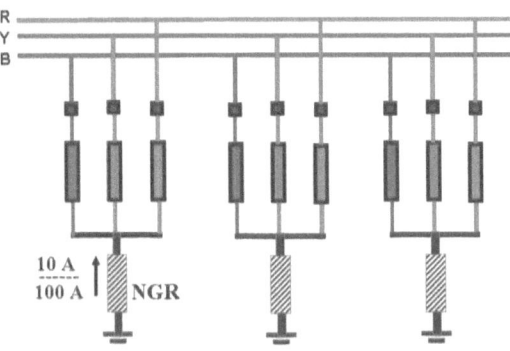

Figure 7.8: Generator on common bus with individual NGR

(a) High resistance grounded system
The ground fault current is limited to within 10 – 15A. All NGRs can be in circuit. Third harmonic current between two machines encounters two NGRs of sufficiently high value. The resulting circulating current is very low.

(b) Low resistance grounded system
The ground fault current is limited to say 100 A. Some prefer to keep only one NGR in circuit at any time. This requires switching device for neutral (neutral breaker/isolator). Some others prefer to keep all NGRs in circuit. If sensitive Restricted Earth Fault (REF) scheme is provided for each unit which can clear internal fault in machine within 100msec, the later alternative is preferred.

(ii) Generators with GTs (Figure 7.9).
In majority of power stations, this scheme is adopted. All NGRs are permanently in circuit. The ground fault current is limited to within 10A. The vector group of Generator Transformer is star – delta. Delta (on generator side) offers zero sequence isolation between individual generator and rest of the system. For earth faults within generator or busduct to transformer, system contribution is blocked due to the presence of delta. Third harmonic current circulation between two machines is not theoretically possible. Any stator earth fault protection provided on generator is inherently REF protection and does not need coordination with ground relays (51N) on system (220kV) side.

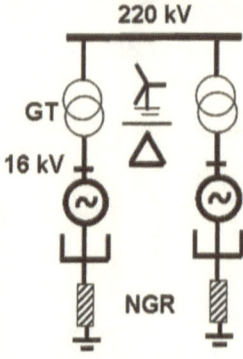

Figure 7.9: Generator with GT

7.5 ZIG ZAG GROUNDING TRANSFORMER COMMON TO ALL THE UNITS

The scheme is shown in Figure 7.10. In some stations generator neutrals are kept floating and the bus is grounded through Zig Zag grounding transformer. The neutral of Zig Zag transformer is grounded through NGR. Operating principle of Zig Zag grounding transformer is given in Sec 5.11.1. As in Figure 7.6, in this case also, the ground fault current supplied by the plant remains almost the same irrespective of number of units operating in parallel. Unlike Figure 7.6, however, zero sequence (third harmonic) current circulation among machines is eliminated.

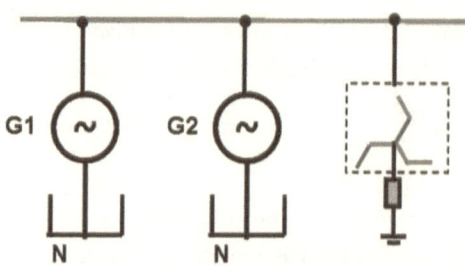

Figure 7.10: Zig Zag Grounding

One practical case to illustrate evolution of this type of grounding is shown in Figure 7.11.

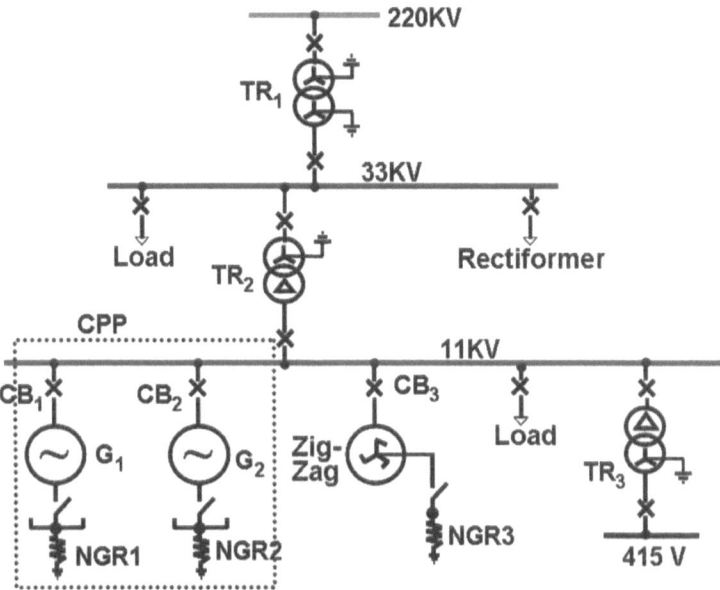

Figure 7.11: Evolution of Grounding

Before the Captive Power Plant (CPP) was commissioned, the plant loads were fed by transformers TR_1, TR_2 and TR_3. Since 11kV system was ungrounded, Zig Zag transformer was installed to ground the bus. After commissioning CPP, following procedure is adopted to bring the units on line:

(i) Run up generators G_1 and G_2 and synchronize using breakers CB_1 and CB_2.
(ii) Close NGR_1 and NGR_2.
(iii) To avoid multiple grounding, manually trip Zig Zag transformer using CB_3.

During parallel operation if CPP units trip, manually close CB_3 to reestablish grounding. If the operator fails to close CB_3 the 11kV system remains ungrounded. To obviate human error, CB_3 is kept always closed and NGR_1 and NGR_2 are permanently kept off.

7.6 LV GENERATORS GROUNDING

415V generators are mostly solidly grounded. To prevent circulation of third harmonic current among the machines, the neutral of only one generator is grounded (Figure 7.12). This offers return path for ground fault current. The neutral isolating device can be a switch, contactor or breaker. The switch has to be manually opened or closed. The contactor or breaker can be remotely opened or closed through control logic depending on the neutral of which machine is to be grounded.

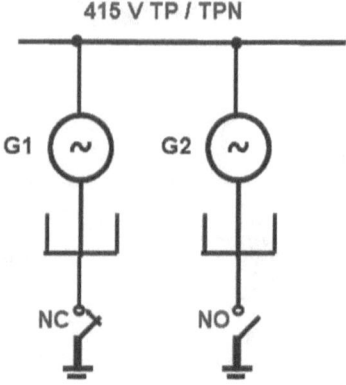

Figure 7.12: LT Generators Grounding

In case of TPN distribution, zero sequence current circulation can be significant even if neutral of only one generator is grounded (Figure 7.13). A separate neutral contactor instead of neutral link is to be provided on the bus side for isolation.

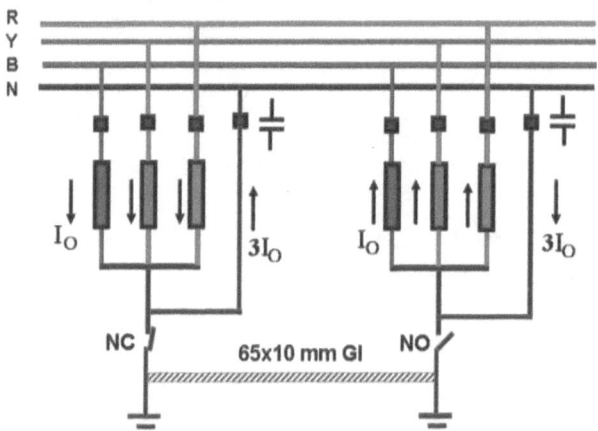

Figure 7.13: LT System Grounding (TPN)

If earthing connection is not proper, even if neutral of only one generator is grounded in TP system, zero sequence current circulation can be significant (Figure 7.14). The connection arrangement from generator neutral to isolator and isolator to station grounding grid has to be physically verified to avoid this type of site problem.

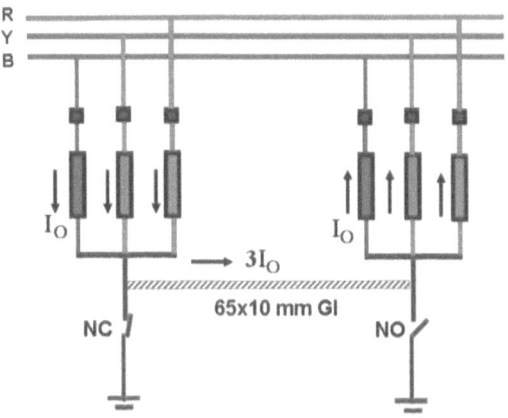

Figure 7.14: LT System Grounding (TP)

7.7 MV GENERATORS GROUNDING

Winding damages in rotating machines are not of serious concern. The repairs can be done by local rewinding agency. However in case of damage to core, repairs can't be carried out at site. The machine has to be sent back to manufacturer's works for repair resulting in prolonged loss of production (Refer Sec 5.8.2). To limit the ground fault current, the neutral of generator is earthed through a resistor. Depending on the value of limiting ground fault, the grounding can be classified as high resistance grounding or low resistance grounding. Salient points have already been discussed in Sec 5.9.1 and 5.9.2.

7.8 HIGH RESISTANCE GROUNDING

A typical sizing calculation of the NGT and NGR for a 600MW unit is explained [9].

7.8.1 Data
- Rated Output: 600MW
- Rated Voltage, V_{LL}: 20kV
- Phase Voltage, $V_{PH} = 20/\sqrt{3} = 11.55$kV
- Capacitances
- Generator Capacitance, $C_G = 0.213\mu F$
- Iso phaseduct Capacitance, $C_{IPB} = 0.01048\mu F$
- Generator Transformer Capacitance, $C_{GT} = 0.012\mu F$
- Unit Transformer Capacitance, $C_{UT} = 0.027\mu F$
- Surge Capacitance, $C_S = 0.125\mu F$
- System Capacitance = $C = C_G + C_{IPB} + C_{GT} + C_{UT} + C_S$
  = $0.38748\ \mu F/phase$

It can be seen that generator winding capacitance and surge capacitance are the dominant factors. In fact surge capacitance value can be as high as 0.25μF in many cases. Hence inaccuracies in the value of bus duct or transformer winding capacitances are not of serious concern.

7.8.2 Calculations

(i) Capacitive Reactance/Phase, $X_{cg} = \dfrac{1}{2\pi f C}$

$= 8214.8 \, \Omega$

(ii) Capacitive charging Current during earth fault, $I_C = \dfrac{3 x V_{ph}}{X_{cg}}$

$= \dfrac{3 \times 11.55 \times 10^3}{8214.8}$

$= 4.21 \, A$

(iii) The primary voltage rating of NGT is selected as 15kV. This is a conservative value as neutral to ground voltage for high resistance grounded system during ground fault will be maximum phase voltage (11.55kV). Refer Sec 4.4.

(iv) The required NGT rating = $V_L \times I_C$

$= 15kV \times 4.21 \, A$

$= 63.2 \, KVA$

(v) As per CEA (Central Electricity Authority) guidelines, the NGT shall be sized for a 5 minute duty. For a 5 min duty, the overload factor for determining the continuous rating is 2.8.

Therefore, continuous rating of NGT = $\dfrac{63.2}{2.8}$

$= 23 \, KVA$

The above calculation is very conservative. Earth faults will be cleared well within 10 secs. As per earlier practices, a factor of 6 for 30 secs duty is adequate.

This will result in a NGT with much less continuous rating.

Continuous rating of NGT = $\dfrac{63.2}{6}$

$= 11 \, KVA$

(vi) The loading resistor (R_L) is so selected that the resistive current is slightly greater than capacitive current. It is ensured by using a safety factor of 1.1.

The resistive current, $I_R = 1.1 \times I_C$

$= 1.1 \times 4.21$

$= 4.63 A$

Required value of resistance, $R'_L = \dfrac{V_{ph}}{I_R}$

$$= \dfrac{11.55 \times 10^3}{4.63}$$

$$= 2493 \, \Omega$$

(vii) Considering NGT ratio as 15/0.24kV, the required value of loading resistor, R_L

Loading resistance, $R_L = \dfrac{R'_L}{(15/0.24)^2}$

$$= 0.638 \Omega$$

Therefore the value of loading resistor R_L is chosen as 0.64Ω.

(viii) For a highly oversized design, the selected rating of NGT is 15/0.24kV, 25KVA, 5 min with a loading resistor of 0.64Ω. For optimum design, the NGT rating of 15/0.24kV, 15KVA, 30 sec with a loading resistor of 0.64Ω will suffice.

(ix) The voltage ratio of NGT should be selected in such a way that the resulting value of loading resistor R_L is not very small. It is recommended that the loading resistor should be preferably greater than 0.5Ω to ensure proper operation of 100% stator earth fault protection with 20Hz voltage injection [10]. To achieve the same, two options are available:

(a) Increased secondary voltage of NGT. 500V can be selected (instead of 240V) as the secondary voltage of NGT.

(b) Reduced primary voltage for NGT. The neutral to ground voltage during earth fault will be equal to phase voltage of generator. The minimum rating of NGT primary winding can be 1.2 to 1.3 times phase voltage of generator. For the given example the NGT primary voltage can be 14-15kV. 15kV is chosen in the above example.

(x) A over dimensioning factor is many times considered in NGT sizing to account for field forcing. The implication of field forcing on NGT sizing is discussed in detail in the next section.

It may be noted the actual fault current will be marginally less than design earth fault current as the following are ignored in the calculation:

(i) Resistance of NGT secondary and connecting cable to resistor
(ii) Leakage reactance of NGT

7.8.3 GT Sizing and Exciter Field Forcing

➤ When sizing NGT, some design guides recommend over dimensioning factor of 1.3 to 1.4 to account for field forcing. Following discussions critically examine the influence of field forcing on NGT sizing.

- Unlimited forcing is typically for 1 second, to allow excitation to force to ceiling voltage for close in faults that are cleared in primary or backup clearing time.
- There is a timer in the Over Excitation limiter logic (OEL) that allows for unlimited forcing typically for 1 second. After unlimited forcing, the IEEE 50.13 curve is used to compute excess heating in the rotor. At a point when curve indicates that increased field current is not allowable, a field current regulator brings the field current rapidly back to full load rating [34].
- Line to line voltage of generator PT is connected to AVR. For any line to ground fault on generator side, line voltage is almost unaffected as high resistance grounded system is provided for generator. In this case field forcing will not happen theoretically.
- In majority of cases, field forcing happens only for grid faults. Consider R phase fault on grid (Refer Figure 7.15). Similar case is already discussed in Sec 6.7A. The line voltages on delta side (generator side) reduce significantly (Figure 7.16). Since the feedback signal to AVR is line voltages, field forcing is initiated immediately. Within 100 msec the fault in EHV system is removed and field voltage/current is correspondingly reduced.

Figure 7.15: Generator and GT Circuit

- From Figure 7.16, during grid fault, it can be seen that the neutral voltage is zero. It also follows intuitively from the fact that, a line to ground on star side of transformer is reflected as line to line fault on delta side of transformer (Figure 7.17).

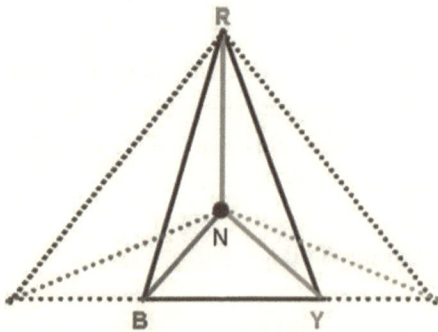

Figure 7.16: Generator Voltage Phasor for Line to Ground Fault in Grid

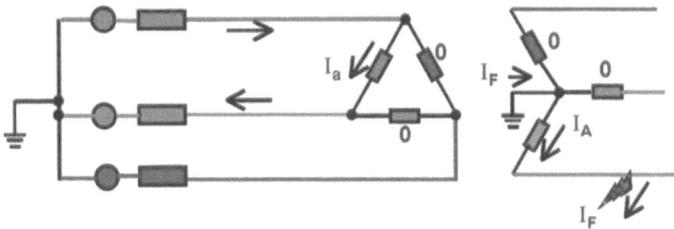

Figure 7.17: Fault Distribution for Line to Ground Fault

From the above, following are summarized:

1. During earth fault on generator side, field forcing will not happen as line voltages on generator side are almost balanced. Stator earth fault protection trips the unit within a second.
2. During internal phase fault on generator side, differential protection will trip the unit within 100 msec.
3. Field forcing happens for external ground fault in grid. However the neutral voltage on generator side is nearly zero during external grid faults.
4. Unlimited field forcing is typically for a maximum of 1 second.
5. Considering the above, over-dimensioning NGT for field forcing condition is not required.

7.9 LOW RESISTANCE GROUNDING

The ground fault current is limited to about 100A to 400A compared to 10A in high resistance grounded system. On a 11kV system, with ground fault current limited to 400A, value of NGR is approximately given by:

$$R_G \cong \frac{\left(11000/\sqrt{3}\right)}{400}$$

$$\cong 16\,\Omega.$$

The resistor is directly connected between neutral and ground (Figure 7.18). Current relay in neutral circuit is possible as ground fault current is not too low.

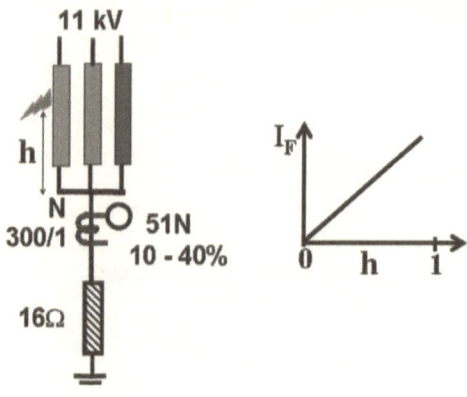

Figure 7.18: Sensitivity of ground fault Protection

7.10 SENSITIVITY OF GROUND FAULT PROTECTION

The variation of fault current with respect to distance from neutral is linear.

For fault on terminal (Figure 7.18),

$$I_F^T = \frac{\left(11000/\sqrt{3}\right)}{16}$$

$\cong 400$ A

For fault at a distance 'h' from neutral,

$$I_F = h \, I_F^T \qquad (7.1)$$

For fault on neutral, h = 0
$I_F = 0$
CT ratio: 300/1
Assume the relay is set for a minimum pick up of 10%.
Minimum fault current for relay pick up:
$I_F = 300 \times 0.1$
= 30 A

From Eqn (7.1):
$30 = h \times 400$
h = 0.075 (7.5%)

For this relay setting, 7.5% of winding from neutral is not protected. If setting is increased, zone of unprotected winding also increases correspondingly.

Compared to high resistance grounded system, the core damage at the faulted location will be more. Typical REF scheme for alternator is shown in Figure 7.19.

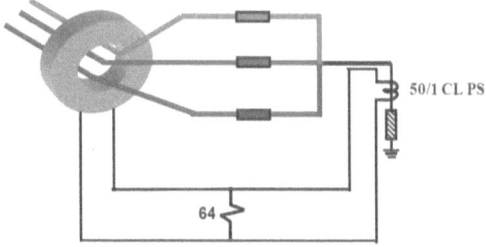

Figure 7.19: REF Protection with CBCT

7.11 HYBRID GROUNDING

It combines the advantage of high resistance grounding (low fault current and less core damage) and low resistance grounding (sufficient fault current and high sensitivity). The scheme [8] is shown in Figure 7.20.

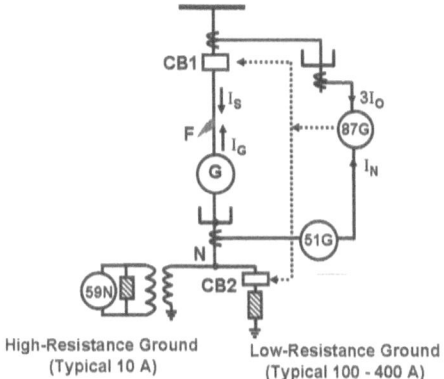

Figure 7.20: Hybrid grounding

For an internal fault at F, after CB1 trips, system contribution I_S will be zero but large generator contribution I_G will continue to flow till the flux in the machine decays to a low value. The decay time constant is large (in the range of 10 sec). The generator is initially grounded through low resistance. For a ground fault, sufficient fault current flows for ground differential 87G to pick up. It trips both CB1 and CB2 within 100 to 300msec. Once CB2 trips, the generator is grounded through high resistance. The generator contribution subsequently is very less during flux decay period.

This method of grounding is limited to only smaller size machines (less than 20MW). It does not find widespread use due to following reasons
- Ground fault current immediately after fault is higher
- Requirement of two NGRs and NGT
- Complex relaying and switching arrangement

7.12 GROUNDING MIX-UP

The fallacy in mixing up grounding is explained with an example (Figure 7.21).

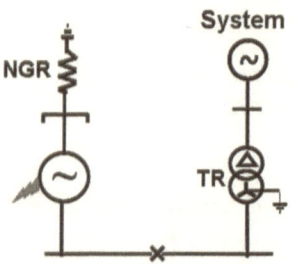

Figure 7.21: Grounding Mix-up

The generator is grounded through NGR. The transformer is solidly grounded. For any ground fault external to the generator, the current fed by it is limited by NGR to say 10A. But for fault within the generator, the current at the point of fault is determined by external system grounding. If external system (in this case transformer TR) is solidly grounded, it can contribute say 40kA. Thus at the point of fault within the generator, the current is not limited to 10A but can be as high as 40kA (Figure 7.22).

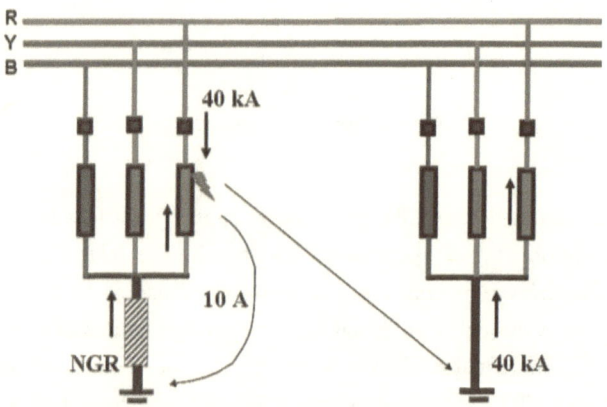

Figure 7.22: Fault contribution

The very purpose of providing NGR to limit core damage in generator is defeated. *This brings out the important fact that when two systems are paralleled, grounding type has to be compatible.*

To avoid grounding mix-up, one alternative is to introduce NGR in transformer (Figure 7.23). Insulation of transformer neutral is usually adequate to withstand rise in neutral voltage during ground faults (Sec 4.4). As long as earth fault protection is provided to isolate the earth faults, existing earthed grade cables can be retained (Sec 5.13). With NGR earth fault currents will be low (100A or less). Depending on CT ratio and value of NGR, earth fault setting (51N) needs to be sensitive (say 1 to 5%). If earth fault setting is too sensitive, it may lead to spurious tripping during charging or transient conditions. To improve security, NDR (Neutral Displacement Relay) is connected to secondary of open delta PT (Sec 4.9). Contact of NDR is wired in series with tripping contact of ground fault relay (51N).

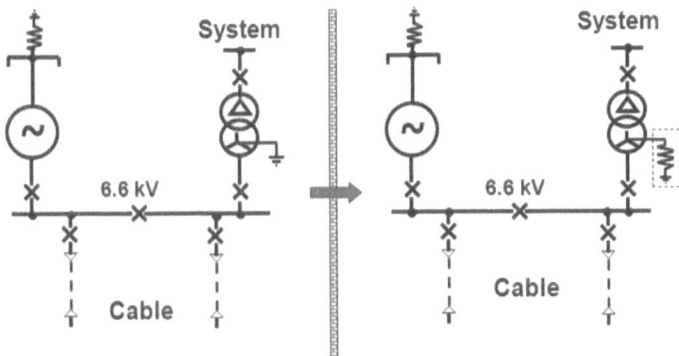

Figure 7.23: Alternative 1

The other alternative is to introduce 1:1 Generator Transformer with vector group of star – delta (Figure 7.24). The star neutral is solidly grounded. The delta winding offers ground fault isolation between generator and the system. The scheme is similar to Figure 7.9 adopted for large units. The economics of providing additional GT needs to be looked into.

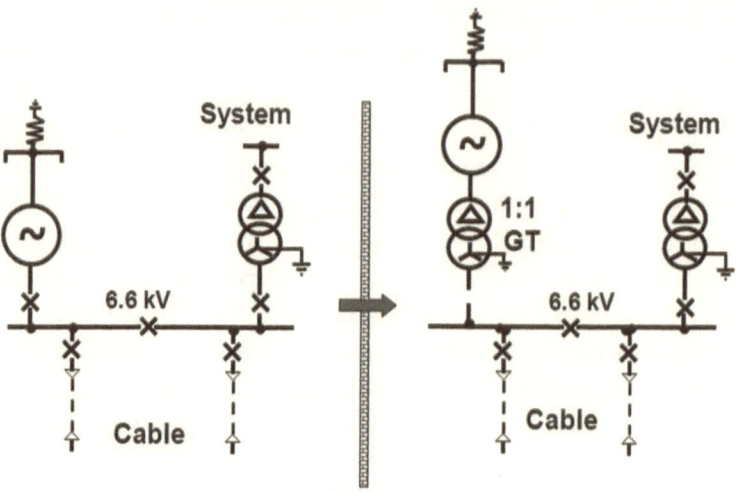

Figure 7.24: Alternative 2

References

Under References, articles, books and standards used by the author are listed. The contents of the book are consolidation and elaboration of topics scattered in various articles written by author in the past. References cited in these articles can be consulted for further reading. The list is not exhaustive. Any omission is unintentional. Unlike in the past, the reader can search on the internet for the particular topic mentioned in the book and get more related information.

1. 'Grounding transformer specification without ambiguity', K Rajamani and H C Mehta, IEEMA Journal, Aug 2001, pp 52-54.
2. 'Peculiarities of delta connection in electrical power systems', K Rajamani, IEEMA Journal, Dec 2003, pp 38-42.
3. 'Earthing of electrical system – Part 1', K Rajamani, IEEMA Journal, Aug 2004, pp 37-40.
4. Earthing of electrical system – Part 2', K Rajamani, IEEMA Journal, May 2005, pp 32-36.
5. 'Grounding of electrical system – Part 1', K Rajamani, IEEMA Journal, May 2006, pp 52-56.
6. 'Grounding of electrical system – Part 2', K Rajamani, IEEMA Journal, June 2006, pp 51-58
7. 'Generator neutral grounding practices', K Rajamani and Bina Mitra, IEEMA Journal, Aug 2007, pp 89-97.
8. 'Evaluation of generator parameters by online testing', K Rajamani and Bina Mitra, IEEMA Journal, Feb 2008, pp 68-82.
9. 'Stator Earth Fault Protection of large generator (95%) – Part 1' K Rajamani and Bina Mitra, IEEMA Journal, May 2013, pp 76-80
10. 'Stator Earth Fault Protection of large generator (100%) – Part 2' K Rajamani and Bina Mitra, IEEMA Journal, May 2013, pp 81-86.
11. 'Zig Zag transformer – Fault current distribution, Short circuit testing and single phase loading', K Rajamani and Bina Mitra, IEEMA Journal, July 2013, pp 84-91.

12. 'Cable sequence impedance measurement at site', K Rajamani and Bina Mitra, IEEMA Journal, August 2013, pp 84-86.
13. 'Application of Capacitors in Electrical Power Systems', K Rajamani and Bina Mitra, Eighth International Conference on Capacitors – CAPCIT 2014, IEEMA November 2014, New Delhi pp 177-183.
14. 'Conceptual Clarifications in Electrical Power Engineering – Part 1' K Rajamani, IEEMA Journal, Aug 2016, pp 69-80.
15. TVE10-07 Prysmian guideline on cable metallic sheath earthing at GIS
16. Earthing principles and practices: R W Ryder
17. Electrical earthing and accident prevention: edited by M G Say
18. Protective Relaying – Principles & Applications: J Lewis Blackburn, Marcel Dekker Pub.
19. 'Symmetrical components for Power Systems Engineering', J Lewis Blackburn, Marcel Dekker, 1993.
20. 'Analysis of faulted power systems', P M Anderson, IEEE Press, 1995.
21. Alstom Network Protection and Automation Guide, 2011
22. IE Rules 1956–2000
23. IS – 3043 (2001): Code of practice for earthing
24. IS 7098-2 (2005): Specification for Cross linked Polyethylene Insulated PVC Sheathed Cables for working voltages from 3.3kV up to and including 33kV.
25. IEEE Std 575 (1988) - IEEE Guide for the Application of Sheath-Bonding Methods for Single-Conductor Cables and the Calculation of Induced Voltages and Currents in Cable Sheaths
26. IEEE C62.92.4 (1991) – IEEE Guide for the Application of Neutral Grounding in Electrical Utility Systems, Part IV—Distribution
27. IEC 60044-2 (2003): Instrument transformers – Inductive voltage transformers
28. IEEE Std 142 (2007): Grounding of Industrial and Commercial Power Systems
29. IEEE Std 80 (2013): IEEE Guide for safety in AC Substation Grounding
30. IEC 60364-1 (2005): Low voltage electrical Installations – Fundamental principles, assessment of general characteristics, definitions
31. IEC 60502-2 (2005): Power Cables with extruded insulation and their accessories for rated voltages from 1kV up to 30kV
32. IEC 60865 -1 (2011): Short Circuit Currents – Calculation effects – Definitions and Calculation methods
33. Bender Technical Bulletin - Digital Ground Fault Monitor/Ground Detector Controller for ground fault location system
34. Generator Over Excitation Capability and Excitation System Limiters', A. Murdoch et al, IEEE PES, 2001, WPM, Columbus, OH
35. IEC 60076-6 (2007): Power transformer – Part 6 – Reactor
36. IEEE Guide C57.13.3 - 2014: Grounding of Instrument Transformer Secondary Circuits and Cases

www.ingramcontent.com/pod-product-compliance
Lightning Source LLC
Chambersburg PA
CBHW020740180526
45163CB00001B/295